Methods in Enzymology

Volume 421
ADVANCED BACTERIAL GENETICS: USE OF TRANSPOSONS AND PHAGE FOR GENOMIC ENGINEERING

METHODS IN ENZYMOLOGY

EDITORS-IN-CHIEF

John N. Abelson   Melvin I. Simon

DIVISION OF BIOLOGY
CALIFORNIA INSTITUTE OF TECHNOLOGY
PASADENA, CALIFORNIA

FOUNDING EDITORS

Sidney P. Colowick and Nathan O. Kaplan

Methods in Enzymology

Volume 421

Advanced Bacterial Genetics: Use of Transposons and Phage for Genomic Engineering

EDITED BY

Kelly T. Hughes

DEPARTMENT OF BIOLOGY
UNIVERSITY OF UTAH
SALT LAKE CITY, UTAH

Stanley R. Maloy

DEPARTMENT OF BIOLOGY
CENTER FOR MICROBIAL STUDIES
SAN DIEGO STATE UNIVERSITY
SAN DIEGO, CALIFORNIA

AMSTERDAM • BOSTON • HEIDELBERG • LONDON
NEW YORK • OXFORD • PARIS • SAN DIEGO
SAN FRANCISCO • SINGAPORE • SYDNEY • TOKYO
Academic Press is an imprint of Elsevier

ELSEVIER

Academic Press is an imprint of Elsevier
525 B Street, Suite 1900, San Diego, California 92101-4495, USA
84 Theobald's Road, London WC1X 8RR, UK

Permissions may be sought directly from Elsevier's Science & Technology Rights
Department in Oxford, UK: phone: (+44) 1865 843830, fax: (+44) 1865 853333,
E-mail: permissions@elsevier.com. You may also complete your request on-line
via the Elsevier homepage (http://elsevier.com), by selecting "Support & Contact"
then "Copyright and Permission" and then "Obtaining Permissions."

For information on all Elsevier Academic Press publications
visit our Web site at www.books.elsevier.com

ISBN-13: 978-0-12-373749-6
ISBN-10: 0-12-373749-4

PRINTED IN THE UNITED STATES OF AMERICA
07 08 09 10 9 8 7 6 5 4 3 2 1

Table of Contents

Section I. Strain Collections and Genetic Nomenclature

Section II. Transposons

Section III. Phage

Contributors to Volume 421

Article numbers are in parentheses following the names of contributors.
Affiliations listed are current.

DAVID N. BALDWIN (10), *Human Biology Division, Fred Hutchinson Cancer Research Center, Seattle, Washington*

DANA BOYD (8), *Department of Microbiology and Molecular Genetics, Harvard Medical School, Boston, Massachusetts*

MIKHAIL BUBUNENKO (15), *Molecular Control and Genetics, National Cancer Institute at Frederick, Frederick, Maryland*

VERONICA CASAS (20), *Center for Microbial Sciences, San Diego State University, San Diego, California*

NINA COSTANTINO (15), *Molecular Control and Genetics, National Cancer Institute at Frederick, Frederick, Maryland*

DONALD L. COURT (15), *Molecular Control and Genetics, National Cancer Institute at Frederick, Frederick, Maryland*

JENNY A. CRAIN (19), *Department of Biology, Center for Microbial Sciences, San Diego State University, San Diego, California*

SIMANTI DATTA (15), *Molecular Control and Genetics, National Cancer Institute at Frederick, Frederick, Maryland*

MICHAEL EHRMANN (8), *Centre for Medical Biotechnology, University Duisburg-Essen, Essen, Germany*

MARKUS ESER (8), *School of Biosciences, Cardiff University, Cardiff, United Kingdom*

ELIORA GACHELET (9), *Department of Microbiology, University of Washington, Seattle, Washington*

LARRY GALLAGHER (12), *Department of Genome Sciences, University of Washington, Seattle, Washington*

NATHALIE GARCIA-RUSSELL (17), *Department of Biology, San Diego State University, San Diego, California*

JEFFREY GARDNER (18), *Department of Microbiology, University of Illinois, Urbana, Illinois*

TANJA HENRICHS (8), *School of Biosciences, Cardiff University, Cardiff, United Kingdom*

DAVID R. HILLYARD (4), *Department of Pathology, University of Utah, Salt Lake City, Utah*

KELLY T. HUGHES (1, 7, 13, 14), *Department of Biology, University of Utah, Salt Lake City, Utah*

JOYCE E. KARLINSEY (14, 16), *Department of Microbiology, University of Washington, Seattle, Washington*

JOHN R. KIRBY (3), *School of Biology, Georgia Institute of Technology, Atlanta, Georgia*

CHANGHAN LEE (14), *Department of Microbiology, University of Washington, Seattle, Washington*

STANLEY R. MALOY (1, 2, 5, 6, 13, 18, 19), *Department of Biology, Center for Microbial Sciences, San Diego State University, San Diego, California*

COLIN MANOIL (12), *Department of Genome Sciences, University of Washington, Seattle, Washington*

SAMANTHA S. ORCHARD (17), *Department of Biology, San Diego State University, San Diego, California*

ELIZABETH RAMAGE (12), *Department of Genome Sciences, University of Washington, Seattle, Washington*

MICHAEL J. REDD (4), *Department of Oncological Science, Huntsman Cancer Institute, University of Utah, Salt Lake City, Utah*

WILLIAM S. REZNIKOFF (11), *Department of Biochemistry, University of Wisconsin-Madison, Madison, Wisconsin*

FOREST ROHWER (20), *Center for Microbial Sciences, San Diego State University, San Diego, California*

NINA R. SALAMA (10), *Human Biology Division, Fred Hutchinson Cancer Research Center, Seattle, Washington*

JAMES A. SAWITZKE (15), *Molecular Control and Genetics, National Cancer Institute at Frederick, Frederick, Maryland*

ANCA M. SEGALL (17), *Center for Microbial Sciences, San Diego State University, San Diego, California*

LYNN C. THOMASON (15), *Molecular Control and Genetics, National Cancer Institute at Frederick, Frederick, Maryland*

BETH TRAXLER (9), *Department of Microbiology, University of Washington, Seattle, Washington*

CHERI TURNER (12), *Department of Genome Sciences, University of Washington, Seattle, Washington*

KELLY M. WINTERBERG (11), *Department of Biology, University of Utah, Salt Lake City, Utah*

CHRISTOPHER WOZNIAK (14), *Department of Microbiology, University of Washington, Seattle, Washington*

PHILIP YOUDERIAN (18), *Department of Biology, Texas A&M University, College Station, Texas*

Preface

Over 70 years ago, microbial genetics led to the molecular revolution in biology, providing powerful new methods for exploring the fundamental mechanisms of life. Beginning in the mid-1970s, the development of molecular cloning coupled with the characterization of transposons led to a second revolution in molecular genetics, providing a set of indispensable tools that facilitated the expansion of basic genetic selections and screens from the *E. coli*, *Bacillus*, and *Salmonella* paradigms throughout the microbial world. Recently, another revolution in molecular genetics has been stimulated by the sequence determination of hundreds of complete genomes in combination with methods to determine RNA and protein expression levels as well as a complete library of sets of interacting proteins and genome-scale three-dimensional protein structure determinations. These new approaches provide insights into how each of the multiple different cell components changes in response to a particular stimulus—a Gestalt perspective called systems biology. However, confirming the predictions from whole genome approaches requires genetic analysis to verify the roles of particular gene products. Combining systems biology and genetic approaches will provide novel insights into basic biological processes and into the role of microbes in the environment, health, and disease.

Transposons and bacteriophage continue to play a key role in the dissection of fundamental processes, whether at the single pathway or whole genome scale. This volume describes some of the advances in bacterial genetics that have expanded the uses of transposons and bacteriophage in genomic engineering. The basics of genetic analysis using transposons are covered in Chapters 1, 2, 3, and 5. Transposons provide efficient methods of constructing mutations and selectable linked genetic markers, moving mutations into new strains, facilitating mutagenesis of defined regions of the genome (Chapter 6), and constructing fusions (Chapter 13). In addition, because multiple copies of a transposon in a cell provide regions of genetic homology that can be placed at any desired position, transposons can be used to select for recombination events that yield defined deletions and duplications (Chapter 7). The ability of transposons to generate large chromosomal duplications allows for the identification of all essential genes in an organism and the isolation of reporter fusions to characterize the expression of these genes (Chapter 4). In addition, transposons have been modified for the construction of specific reporters or the introduction of small peptides for the probing of three-dimensional structure and protein location or topology in the cell (Chapters 9 and 12). Uses of transposons to tag specific proteins with protease sites or with antigen sites are useful for controlling and measuring levels of given proteins under any condition (Chapters 8, 9, and 12).

With the acquisition of whole genome sequences, it is now possible to characterize transposon mutant libraries on a genomic scale. Methods are described for creating and maintaining such libraries as well as combining genomic microarray analysis to study the impact of the transposon insertions on whole genome gene expression under any set of conditions desired (Chapters 10, 11, and 12). Transposons can also be utilized in combination with inducible promoters to probe whole genomes for genes that, when induced, have positive or negative regulatory effects on the system under investigation (Chapter 14).

New advances in bacteriophage technologies that have virtually eliminated the artifacts of working with multicopy plasmid vectors are described. Phage recombination systems have been harnessed to allow the use of synthetic oligonucleotides for engineering mutations or fusions for the study of gene regulation, protein topology, or cell biology (Chapters 15 and 16). Phage systems have been modified that allow the characterization of chromosome structure (Chapter 17), for studying protein–DNA, protein–protein, and protein–RNA interactions (Chapter 18), and for the manipulation of specific segments of the bacterial chromosome (Chapter 19). Advances in phage metagenomics (Chapter 20) allow the identification of millions of novel proteins from diverse environments around the globe, whether or not the phage or host can be grown in the lab (Chapter 20).

In short, microbial genetics remains a vibrant and important field that has both been enriched by new methods of systems biology and provides useful tools for systems biology. Transposons and phage continue to provide invaluable new tools for dissecting the structure, function, regulation, and physiology of microbes.

We would like to thank the authors of this volume for their contributions. We are especially grateful for the project management assistance of Cindy Minor and Jamey Stegmaier for their patience and persistence.

KELLY T. HUGHES
STANLEY R. MALOY

METHODS IN ENZYMOLOGY

VOLUME 244. Proteolytic Enzymes: Serine and Cysteine Peptidases
Edited by ALAN J. BARRETT

VOLUME 245. Extracellular Matrix Components
Edited by E. RUOSLAHTI AND E. ENGVALL

VOLUME 246. Biochemical Spectroscopy
Edited by KENNETH SAUER

VOLUME 247. Neoglycoconjugates (Part B: Biomedical Applications)
Edited by Y. C. LEE AND REIKO T. LEE

VOLUME 248. Proteolytic Enzymes: Aspartic and Metallo Peptidases
Edited by ALAN J. BARRETT

VOLUME 249. Enzyme Kinetics and Mechanism (Part D: Developments in Enzyme Dynamics)
Edited by DANIEL L. PURICH

VOLUME 250. Lipid Modifications of Proteins
Edited by PATRICK J. CASEY AND JANICE E. BUSS

VOLUME 251. Biothiols (Part A: Monothiols and Dithiols, Protein Thiols, and Thiyl Radicals)
Edited by LESTER PACKER

VOLUME 252. Biothiols (Part B: Glutathione and Thioredoxin; Thiols in Signal Transduction and Gene Regulation)
Edited by LESTER PACKER

VOLUME 253. Adhesion of Microbial Pathogens
Edited by RON J. DOYLE AND ITZHAK OFEK

VOLUME 254. Oncogene Techniques
Edited by PETER K. VOGT AND INDER M. VERMA

VOLUME 255. Small GTPases and Their Regulators (Part A: Ras Family)
Edited by W. E. BALCH, CHANNING J. DER, AND ALAN HALL

VOLUME 256. Small GTPases and Their Regulators (Part B: Rho Family)
Edited by W. E. BALCH, CHANNING J. DER, AND ALAN HALL

VOLUME 257. Small GTPases and Their Regulators (Part C: Proteins Involved in Transport)
Edited by W. E. BALCH, CHANNING J. DER, AND ALAN HALL

VOLUME 258. Redox-Active Amino Acids in Biology
Edited by JUDITH P. KLINMAN

VOLUME 259. Energetics of Biological Macromolecules
Edited by MICHAEL L. JOHNSON AND GARY K. ACKERS

VOLUME 260. Mitochondrial Biogenesis and Genetics (Part A)
Edited by GIUSEPPE M. ATTARDI AND ANNE CHOMYN

VOLUME 261. Nuclear Magnetic Resonance and Nucleic Acids
Edited by THOMAS L. JAMES

Section I

Strain Collections and Genetic Nomenclature

[1] Strain Collections and Genetic Nomenclature

By STANLEY R. MALOY and KELLY T. HUGHES

Abstract

The ease of rapidly accumulating a large number of mutants requires careful bookkeeping to avoid confusing one mutant with another. Each mutant constructed should be assigned a strain number. Strain numbers usually consist of two to three capital letters designating the lab where they were constructed and a serial numbering of the strains in a central laboratory collection. Every mutation should be assigned a name that corresponds to a particular gene or phenotype, and an allele number that identifies each specific isolate. When available for a particular group of bacteria, genetic stock centers are the ultimate resources for gene names and allele numbers. Examples include the *Salmonella* Genetic Stock Centre (http://www.ucalgary.ca/~kesander/), and the *E. coli* Genetic Stock Center (http://cgsc.biology.yale.edu/). It is also important to indicate how the strain was constructed, the parental (recipient) strain, and the source of any donor DNA transferred into the recipient strain (Maloy *et al.*, 1996).

Introduction

Through the 1960s, genetic nomenclature was a virtual "Tower of Babel." Due to the absence of clear rules for naming genes, each investigator assigned new names based on the method of isolation, which often resulted in the same name being applied to different genes or different names being applied to the same gene. To further confuse the issues, different investigators would each assign allele numbers independently, so two different alleles might have the exact same designation. To eliminate the resulting confusion, Demerec *et al.* (1966, 1968) developed a standard nomenclature for bacterial genes. With the development of new genetic and molecular tools, some modifications have been developed to describe particular types of mutations. With the increasing ease of determining the DNA sequence of mutants, it has become commonplace to simply indicate the amino acid sequence change of an encoded protein rather than assigning an allele number. However, even when the DNA sequence of a mutation is known, a specific allele number is invaluable for tracking the history of a strain and for maintaining large strain collections. The basic rules are described next.

METHODS IN ENZYMOLOGY, VOL. 421
Copyright 2007, Elsevier Inc. All rights reserved.
0076-6879/07 $35.00
DOI: 10.1016/S0076-6879(06)21001-2

Genotype

Genes

Each gene is assigned a three-letter designation, usually an abbreviation for the pathway or the phenotype of mutants. When the genotype is indicated, the three-letter designation is written in lowercase. Multiple genes that affect the same pathway are distinguished by a capital letter following the three-letter designation. For example, mutations affecting pyrimidine biosynthesis are designated *pyr*; the *pyrC* gene encodes the enzyme dihydroorotase, and the *pyrD* gene encodes the enzyme dihydroorotate dehydrogenase. There is only one gene required for the DNA ligase function, so mutations affecting this function are simply indicated *lig*. Three-letter–only designations are also used to indicate mutations such as deletions that affect multiple genes within a multigene operon.

Allele Numbers

Each mutation in the pathway is consecutively assigned a unique allele number. Even multiple mutations constructed by directed mutagenesis are assigned different allele numbers to indicate that they arose independently. A separate sequential series of allele numbers is used for each three-letter locus designation. Blocks of allele numbers are assigned to laboratories by the appropriate genetic stock center. Allele numbers should be used sequentially and carefully monitored to ensure that two different mutations are not named with the same allele numbers.

For example, *pyrC19* refers to a particular *pyr* mutation that affects the *pyrC* gene. In order to distinguish each mutation, no other *pyr* mutation, regardless of the gene affected, will be assigned the allele number *19*. The entire genotype is italicized or underlined (e.g., *pyrC19*). A separate series of allele numbers is used for each three-letter locus designation. In cases where there is only a single gene in a pathway or the particular gene in the pathway is unknown, and hence there is no capital letter following the three-letter symbol, insert a dash before the allele number. For example, *lig-131* refers to a particular mutation in the *lig* gene; *pyr-67* refers to a particular mutation that disrupts the pyrimidine biosynthesis pathway, but it is not yet known which gene in this pathway is mutated.

Insertions

Transposable elements or suicide plasmids can insert in known genes or in a site on the chromosome where no gene is yet known. When an insertion is in a known gene, the mutation is given a three-letter designation, gene designation, and allele number as just described, followed by a double colon, and then the type of insertion element. *Do not* leave blank spaces between the letters or numbers and the colon. For example, a particular Tn*10* insertion within the *pyrC* gene (mutant allele number *103)* may be designated *pyrC103*::Tn*10*.

When a transposon insertion is not in a known gene, it is named according to the map position of the insertion on the chromosome. Such insertions are named with a three-letter symbol starting with *z*. The second and third letters indicate the approximate map position in minutes: the second letter corresponds to 10-minute intervals of the genetic map numbered clockwise from minute 0 (*a* = 0–9, *b* = 10–19, *c* = 20–29, etc.); the third letter corresponds to minutes within any 10-minute segment (*a* = 0, *b* = 1, *c* = 2, etc.). For example, a Tn*10* insertion located near *pyrC* at 23 min is designated *zcd*::Tn*10*. Allele numbers are assigned sequentially to such insertions regardless of the letters appearing in the second and third positions, so that if more refined mapping data suggests a new three-letter symbol, the allele number of the insertion mutation is retained. This nomenclature uses *zaa* (0 min) to *zjj* (99 min). The map position for a given insertion might change with refined mapping resulting in letter changes (i.e., *zae* to *zaf*), but the allele number never does. It is the allele number that defines a particular mutation. Insertion mutations on extrachromosomal elements are designated with *zz*, followed by a letter denoting the element used. For example, *zzf* is used for insertion mutations on an F' plasmid. Insertions with an unknown location are designated *zxx*. Allele designation of insertion mutants in unknown genes based on chromosome map location:

zaa	= insertion at 0–1 min
zab	= insertion at 1–2 min
zac	= insertion at 2–3 min
zad	= insertion at 3–4 min
zae	= insertion at 4–5 min
zaf	= insertion at 5–6 min
zag	= insertion at 6–7 min
zah	= insertion at 7–8 min
zai	= insertion at 8–9 min
zaj	= insertion at 9–10 min
zaa–zaj	= insertion in 0–10 min region
zba–zbj	= insertion in 10–20 min region
zca–zcj	= insertion in 20–30 min region
zda–zdj	= insertion in 30–40 min region
zea–zej	= insertion in 40–50 min region
zfa–zfj	= insertion in 50–60 min region

zga–zgj = insertion in 60–70 min region
zha–zhj = insertion in 70–80 min region
zia–zij = insertion in 80–90 min region
zja–zjj = insertion in 90–100 min region
zxx   = insertion with unknown location
zzf   = insertion on F–plasmid

A few commonly used minitransposon derivatives are designated as follows:

Tn*10*dTet = Tet resistance, deleted for Tn*10* transposase
Tn*10*dCam = Derived from Tn*10*dTet, Cam resistance substituted for Tet resistance
Tn*10*dKan = Derived from Tn*10*dTet, Kan resistance substituted for Tet resistance
Tn*10*dGen = Derived from Tn*10*dTet, Gen resistance substituted for Tet resistance
MudJ   = Kan resistance, forms *lac* operon fusions, deleted for Mu transposase
MudJ-Cam = Derived from MudJ, Cam resistance marker disrupts Kan resistance
MudCam   = Cam resistance substitution between ends of Mu

Plasmids

When writing the genotype of a strain, plasmids are often indicated by a slash (/) after the chromosome genotype. It is important to keep track of the name of the plasmid, the plasmid origin, and the relevant genotype or phenotype carried by the plasmid.

Insertions of suicide plasmids into the chromosome can be indicated as described for transposons. If a duplication is generated it can be described as indicated under chromosomal rearrangements.

Phages

Prophages or plasmids integrated into an attachment site can be indicated by the name of the attachment site followed by a double colon and the phage genotype indicated in brackets. An example is *att*::[P22 *mnt*::Kan].

Chromosome Rearrangements

Chromosome rearrangements including deletions, duplications, and inversions should be indicated by a three-letter symbol indicating the type of rearrangement, followed by the genes involved indicated in parentheses, and then the allele number (Schmid and Roth, 1983). The genes and allele number should be italicized or underlined. Rules for this nomenclature are summarized below.

Deletions   = DEL(genes)allele number
Inversions   = INV(join point gene #1 – join point gene #2)allele number
Duplications = DUP(gene #1*join point*gene #2)allele number
Unknown   = CRRallele number

Phenotype

Growth Phenotypes

It is often necessary to distinguish the phenotype of a strain from its genotype (Maloy *et al.*, 1994). The phenotype is usually indicated with the same three-letter designation as the genotype, but phenotypes start with capital letters and are not italicized or underlined. (For example, strain TR251 [*hisC527 cysA1349 supD*] has a Cys(+) His$^+$ phenotype because the *supD* mutation suppresses the amber mutations in both the *cysA* and the *hisC* genes.)

Antibiotic Resistance

Both two- and three-letter designations are commonly used for antibiotic resistance markers. Both are acceptable, but be consistent. Resistance and sensitivity are indicated with a superscript. Common designations are listed below.

Ap Amp = Ampicillin
Cm Cam = Chloramphenicol
Gm Gen = Gentamicin
Km Kan = Kanamycin
Nm Neo = Neomycin
Sp Spc   = Spectinomycin
Sm Str   = Streptomycin
Tc Tet   = Tetracycline
XG   = X-gal
XP   = X-phosphate
Zm Zeo = Zeomycin

Conditional Alleles

Conditional alleles are indicated by the genotype and allele number followed by the two-letter designation for the conditional phenotypes in parentheses. An example is *leuA414*(Am). Because the two-letter designation is a phenotype, it begins with a capital letter. Note that Ts usually refers to heat-sensitive mutants. Common designations for conditional phenotypes follow:

(Ts) = Temperature-sensitive mutation
(Cs) = Cold-sensitive mutation
(Am) = Amber mutation
(Op) = Opal mutation
(Oc) = Ochre mutation

References

Demerec, M., Adelberg, E., Clark, A., and Hartman, P. (1966). A proposal for a uniform nomenclature in bacterial genetics. *Genetics* **54,** 61–76.

Demerec, M., Adelberg, E., Clark, A., and Hartman, P. (1968). A proposal for a uniform nomenclature in bacterial genetics. *J. Gen. Microbiol.* **50,** 1–14.

Maloy, S., Cronan, J., and Friefelder, D. (1994). "Microbial Genetics," 2nd ed. Jones and Bartlett Publishers, Boston.

Maloy, S., Stewart, V., and Taylor, R. (1996). "Genetic Analysis of Pathogenic Bacteria." Cold Spring Harbor Laboratory Press, NY.

Schmid, M. B., and Roth, J. R. (1983). Genetic methods for analysis and manipulation of inversion mutations in bacteria. *Genetics* **105,** 517–537.

Section II

Transposons

[2]   Use of Antibiotic-Resistant Transposons for Mutagenesis

By STANLEY R. MALOY

Abstract

One of the greatest advances in molecular genetics has been the application of selectable transposons in molecular biology. After 30 years of use in microbial genetics studies, transposons remain indispensable tools for the generation of null alleles tagged with selectable markers, genetic mapping, manipulation of chromosomes, and generation of various fusion derivatives. The number and uses of transposons as molecular tools continues to expand into new fields such as genome sciences and molecular pathogenesis. This chapter outlines some of the many uses of transposons for molecular genetic analysis and strategies for their use.

Introduction

Antibiotic-resistant transposons have revolutionized bacterial genetics. Transposons provide efficient methods of constructing mutations and selectable linked genetic markers, moving mutations into new strains, facilitating localized mutagenesis of defined regions of the genome (see Chapter 6), and constructing fusions (see Chapter 13). In addition, because multiple copies of a transposon in a cell provide regions of genetic homology that can be placed at any desired position, transposons can be used to select for recombination events that yield defined deletions and duplications (see Chapter 7), insertions, or inversions (Kleckner *et al.*, 1977). A brief list of some of the most common uses of transposons follows (modified from Kleckner *et al.*, 1991).

1. Transposons can be inserted at a large number of sites on the bacterial chromosome. It is possible to find transposon insertions in or near any gene of interest.
2. With very rare exceptions, transposon insertions result in complete loss of function.
3. The insertion mutation is completely linked to the phenotype of the transposon in genetic crosses. This makes it easy to transfer mutations into new strain backgrounds simply by selecting for transposon-associated antibiotic resistance.

METHODS IN ENZYMOLOGY, VOL. 421   0076-6879/07 $35.00
Copyright 2007, Elsevier Inc. All rights reserved.   DOI: 10.1016/S0076-6879(06)21002-4

4. It is feasible to screen for transposon insertion mutants after low-level mutagenesis because each mutant that inherits the transposon-associated antibiotic resistance phenotype is likely to have only one insertion. The antibiotic resistance provides a selectable marker for backcrosses to establish that the mutant phenotype is due exclusively to one particular transposon insertion.

5. Transposon insertions in operons are nearly always strongly polar on expression of downstream genes. Thus, transposon insertions can be used to determine whether genes are in an operon.

6. Transposons can generate deletions nearby. This provides a convenient method for isolating adjacent deletion mutations.

7. Transposons can provide a portable region of homology for genetic recombination. Transposon insertions can be used to construct deletions or duplications with defined endpoints, or can serve as sites of integration of other genetic elements.

8. When used as a recipient in a genetic cross, transposon insertions behave as point mutations in fine-structure genetic mapping.

9. Transposon insertions can be obtained that are *near* but not *within* a gene of interest. Such insertions are useful for constructing defined deletions and duplications, as well as for genetic mapping.

10. Special transposons can be used to construct operon or gene fusions.

Transposase

The enzymes that catalyze transposition differ with respect to activity and site selectivity (Craig *et al.*, 2002). Transposase from the commonly used transposons Tn5 and Tn10 show clear DNA sequence specificity for the insertion site, but the site recognized is sufficiently permissive to allow the isolation of insertions in essentially any gene. The transposase from Mu also demonstrates less DNA sequence specificity for the insertion site. Transposase mutants can reduce the sequence specificity of insertion. For example, mutations in the Tn10 transposase gene *tpnA* have been isolated with relaxed site specificity (Bender and Kleckner, 1992). Use of transposase derivatives with altered site specificity can increase the site saturation of transposon insertions, and are particularly useful for isolation of rare mutants.

In addition, a variety of transposons have been engineered to remove the transposase from the transposon, demanding that the transposase is provided from a location adjacent to the transposon or in trans from another transposon or plasmid. Such defective mini-transposons are very useful because once the insertion has been separated from the transposase, the insertions are stable and secondary transposon events are eliminated. These mini-transposons must contain the ends of the transposon required for transposition, typically flanking a useful antibiotic-resistance gene.

Delivery of Transposons

To be used as effective genetic tools, transposons have to be efficiently delivered to the recipient cells. Transposition typically occurs at a low frequency *in vivo*. Therefore, it is essential to have an efficient genetic selection to isolate a collection of transposon insertions in a host. A good delivery system provides a selection for transposition. Various approaches are commonly used for the *in vivo* delivery of transposons.

Phage Delivery Systems

Phage delivery systems take advantage of a transposon insertion on a phage that is unable to lyse or lysogenize the recipient cell. For example, lambda cI::Tn10 P(Am) cannot form lysogens because the Tn10 insertion disrupts the cI gene, and cannot grow lytically in a sup^o recipient because the P gene product is required for phage replication. A lysate of this phage is prepared on an *Escherichia coli* amber suppressor mutant, and then an *E. coli sup^o* recipient is infected with the phage, selecting for tetracycline resistance (Tet^R) encoded by the Tn10. The resulting Tet^R colonies result from transposition of Tn10 from the disabled phage onto the chromosome.

An analogous approach relies on the transfer of a phage carrying a transposon from one bacterial species where the phage can reproduce to a species where the phage can infect but cannot replicate. For example, phage P1 can efficiently infect and replicate in *E. coli*, and can infect *Myxococcus xanthus* but cannot replicate in *M. xanthus*. A lysate of P1 carrying a transposon is grown in *E. coli*, and then this lysate is used to infect *M. xanthus*. When *M. xanthus* is infected with a lysate of P1::Tn5 with selection for kanamycin resistance (Kan^R) encoded by the Tn5, the resulting Kan^R colonies are results of the transposition of Tn5 from the disabled phage onto the chromosome.

Plasmid Delivery Systems

Plasmid delivery systems take advantage of a transposon insertion on a plasmid that is unable to replicate in the recipient cell (i.e., the transposon is carried on a suicide plasmid). The plasmid can be transferred from the permissive host to the nonpermissive host by conjugation, transformation, or electroporation, with selection for an antibiotic resistance encoded by the transposon.

Overexpression of Transposase in trans

The frequency of transposition can often be improved by increasing the concentration of transposase in the cell. For example, the transposase gene(s) can be cloned from a transposon into a vector that places their expression under the control of an easily regulated promoter, such as the *tac* promoter or the *ara* promoter. When the transposon is introduced into a cell producing high levels of transposase, the frequency of transposition is sufficiently high that it can often be detected without a direct selection.

Transposon Pools

It is often useful to isolate a collection of transposon insertions at many different sites in the genome of a bacterium. A population of cells that each contains transposon insertions at random positions in the genome (a transposon "pool") can be used to select or screen for those cells with insertions in or near a particular gene. For a genome the size of *E. coli* or *Salmonella* with about 3000 nonessential genes, a transposon pool made up of approximately 15,000 random insertions will represent an insertion in most nonessential genes. About 1 in 3000 of these cells will contain a transposon insertion in any particular nonessential gene, and about 1 in 100 of these cells will contain a transposon insertion within 1 centisome (i.e., 1 min or 1% of the chromosome length) of any particular gene.

A transposon pool can be made by any transposon delivery system. Figure 1 shows how a Tn10 pool could be constructed in *Salmonella*.

To isolate insertions in or near a gene, cells with transposon insertions are first selected by plating onto a rich medium containing an antibiotic resistance expressed by the transposon. The resulting colonies can then be screened for insertion mutations in a gene or insertions near a gene. An example of how to isolate insertions near a gene is shown in Fig. 2.

cis Complementation

Many transposases act preferentially on the DNA from which they are expressed. Often they must be expressed at high levels to work effectively in transposition. Transitory cis complementation is a mechanism that allows cis complementation for transposition of transposons deleted for their cognate transposase, but after transposition, the DNA encoding transposase is degraded leaving insertions that are not capable of further transposition. This is accomplished by placing the transposase gene

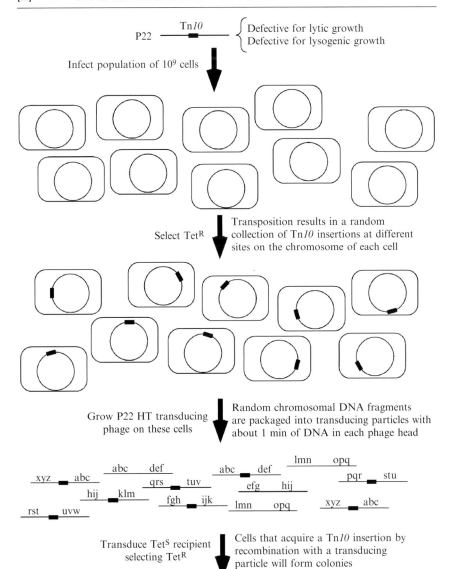

FIG. 1. Isolation of a random pool of Tn*10* insertion mutants.

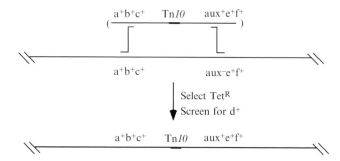

Example screening for a linked Tn*10* insertion linked to *aux*⁺ by replica plating

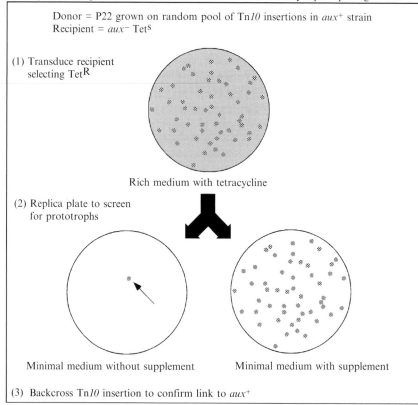

Donor = P22 grown on random pool of Tn*10* insertions in *aux*⁺ strain
Recipient = *aux*⁻ Tet^S

(1) Transduce recipient
    selecting Tet^R

Rich medium with tetracycline

(2) Replica plate to screen
    for prototrophs

Minimal medium without supplement   Minimal medium with supplement

(3) Backcross Tn*10* insertion to confirm link to *aux*⁺

Fig. 2. Screen for Tn*10* insertions near a biosynthetic gene using a generalized transducing phage grown on a random pool of transposon insertions. In the example, *aux*⁺ is prototrophic and can thus grow on minimal medium, while *aux*⁻ is auxotrophic and cannot grow on minimal medium both *aux*⁺ and *aux*⁻ can grow on rich medium.

adjacent to its transposon target under conditions where the transposase is repressed. Introduction of the DNA by transduction, transformation, or electroporation into recipient cells allows for the induction of transposase that can then act on the adjacent transposon for transposition in the recipient cell. At some point, the remaining DNA including the transposase gene is degraded by nucleases resulting in a transposon insertion that is incapable of further transposition.

References

Bender, J., and Kleckner, N. (1992). Is10 transposase mutants that specifically alter target site specificity. *EMBO J.* **11,** 741–750.

Craig, N., Craigie, R., Gellert, M., and Lambowitz, A. (2002). "Mobile DNA II." ASM Press, Washington, DC.

Kleckner, N., Roth, J., and Botstein, D (1977). Genetic engineering *in vivo* using translocatable drug-resistance elements. New methods in bacterial genetics. *J. Mol. Biol.* **116,** 125–159.

Kleckner, N., Bender, J., and Gottesman, S. (1991). Uses of transposons with emphasis on Tn*10*. *Methods Enzymol.* **204,** 139–180.

[3] In Vivo Mutagenesis Using EZ-Tn5™

By JOHN R. KIRBY

Abstract

Epicentre Biotechnologies has developed a suite of transposon-based tools for use in modern bacterial genetics. This chapter highlights the EZ-Tn5™ Transposome™ system and focuses on *in vivo* mutagenesis and subsequent rescue cloning. Many other applications and variations have been described and are available through Epicentre's website at http://www.epibio.com/.

Introduction

The EZ-Tn5™ Transposome™ system from Epicentre provides a rapid and straightforward method for *in vivo* mutagenesis and target identification following rescue cloning from the desired mutant. The EZ-Tn5™ system is based on the hyperactive Tn5 system previously described by Goryshin and Reznikoff (1998).

METHODS IN ENZYMOLOGY, VOL. 421   0076-6879/07 $35.00
DOI: 10.1016/S0076-6879(06)21003-6

The EZ-Tn5 Transposome contains the transposase (Tnp) covalently linked to the 19-bp inverted repeat mosaic ends (ME) that define the transposon, and which flank a conditional origin of replication (R6Kγori) and a Kan^R kanamycin resistance gene. The Transposome is stable enough to be electroporated into any cell for which an electroporation protocol has been established. No host factors are required, thereby making the EZ-Tn5 system useful in many organisms for which genetic systems have yet to be fully developed.

The use of EZ-Tn5 for random mutagenesis has been reported for over 30 organisms, including many Gram-negative and Gram-positive bacteria, and has even been used in the yeast, *Saccharomyces cerevisiae* (Goryshin *et al.*, 2000). The modified EZ-Tn5 transposase inserts randomly and at a frequency approximately 1000 times greater than the wild-type Tn5 transposon (Goryshin and Reznikoff, 1998). Following electroporation, the transposase is activated by Mg^{2+} present in the cell cytoplasm, thereby allowing for a transposition event.

Insertion mutants are selected by plating on media containing kanamycin. Chromosomal DNA can be isolated from the Kan^R mutants with desirable phenotypes, digested with a suitable restriction endonuclease that cuts once in the EZ-Tn5 transposon and once in the chromosome, and then ligated. Only molecules that carry the EZ-Tn5 R6Kγori and flanking chromosomal DNA can replicate in a *pir*$^+$ cloning strain. Plasmids generated in this manner can be sequenced using primers specific to the EZ-Tn5 transposon that direct the sequencing reaction into the chromosomal DNA.

Protocol

The EZ-Tn5^{TM} Transposome^{TM} kit system from Epicentre is used for random insertion of transposable elements into chromosomal DNA. Inheritance of the EZ-Tn5 element can be selected by kanamycin resistance. To identify the site of insertion, a chromosomal restriction fragment containing the kan resistance gene, R6Kγori, and chromosomal sequence, is rescued by cloning.

The chromosomal restriction fragment is rescued by cutting with the appropriate restriction enzyme, recircularized by ligation and transformed into a either a *pir*$^+$ or *pir*-116 cloning strain that provides the π factor *in trans* and allows the fragment to replicate autonomously as a plasmid.

The *pir*$^+$ and *pir*-116 cloning strains (also available through Epicentre) support replication of the plasmid at different copy numbers. The *pir*-116 strain maintains the plasmid at a very high copy number, which facilitates DNA purification, but can lead to toxicity in certain instances. The *pir*$^+$

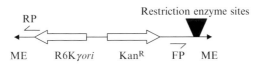

FIG. 1. EZ-Tn5 <R6Kγori/KAN-2> Transposon.

strain maintains the plasmid at lower levels. Once purified, the plasmid can be sequenced using primers specific to EZ-Tn5 that outwardly direct sequencing reactions across the transposon–chromosome junction, allowing for identification of the site of insertion (Fig. 1).

Mutagenesis

1. Electroporate electrocompetent target cells with 1 μl of EZ-Tn5 <R6Kγori/KAN-2>Tnp Transposome. The electrocompetent cells should have a transformation efficiency of >10^7 cfu/μg of DNA, but use cells of the highest transformation efficiency possible to maximize the number of transposon insertion clones.

2. Immediately recover the electroporated cells in appropriate medium. For *Escherichia coli*, add SOC medium to the electroporation cuvette to 1 ml final volume immediately after electroporation. Gently mix cells by pipetting. Transfer to a new tube and incubate at 37° with shaking for 30 to 60 min.

3. If working with *E. coli*, generate a series of dilutions of the recovered cells (e.g., 1:10 and 1:100). Plate 100 μl of each dilution separately onto plates containing 50 μg/ml of kanamycin. Other species may require plating cells on lower concentrations of kanamycin (e.g., 10 to 25 μg/ml).

The number of Kan^R colonies/μl of EZ-Tn5 <R6Kγori/KAN-2> Tnp Transposome will depend on the transformation efficiency of the cells used and the level of expression of the kanamycin resistance marker in that species. Select or screen for insertion mutants of interest by any of a number of methods including observation of a desired phenotypic change or selection for a required function.

Rescue Cloning

1. Prepare genomic DNA from your clone of interest using standard or kit-based techniques.

2. Digest 1 μg of genomic DNA using a restriction endonuclease that cuts near the end of the EZ-Tn5 transposon, but avoids restriction of the

kanamycin resistance gene, the R6Kγ*ori*, or the region between these two elements. Several restriction sites are available and allow for optimization of clone size dependent on the GC content of the target genome.

3. Ligate the fragmented genomic DNA using 0.1 to 1 μmole of DNA and two units of T4 DNA ligase in a standard reaction for 10 min at room temperature. Terminate the ligation reaction and inactivate the T4 DNA Ligase by heating at 70° for 10 min. Other methods such as drop dialysis are also effective.

4. Electroporate electrocompetent *E. coli pir*+ or *pir*-116 cells (*E. coli* expressing the π protein, e.g., TransforMax EC100D *pir*(+) or Transfor-Max EC100D *pir*-116 Electrocompetent *E. coli* from Epicentre) using 1 to 2 μl of the ligation mix.

5. Recover the electroporated cells by adding SOC medium to the electroporation cuvette to 1 ml of final volume immediately after electroporation. Gently mix using a pipette. Transfer to a new tube and incubate at 37° with shaking for 30 to 60 min.

6. Plate cells on LB agar containing 50 μg/ml of kanamycin. Select Kan^R colonies overnight.

7. Miniprep plasmid DNA using standard techniques.

8. Sequence the rescued clones using the forward and reverse EZ-Tn5 <R6Kγ*ori*/KAN-2> Transposon-specific primers (RP and FP) supplied in the kit system from Epicentre (Fig. 2).

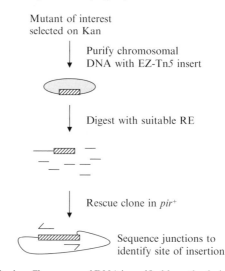

Fig. 2. Rescue cloning. Chromosomal DNA is purified from the desired mutant, digested with a suitable restriction enzyme, ligated and transformed into a *pir*+ strain thereby selecting for plasmids containing the R6Kγ*ori*. The site of insertion is identified by sequencing the junctions of the transposon and flanking chromosomal DNA. Striped box represents the transposon. Leftward and rightward arrows represent primers for sequencing transposon-chromosome junctions.

References

Goryshin, I. Y., and Reznikoff, W. S. (1998). Tn5 *in vitro* transposition. *J. Biol. Chem.* **273,** 7367–7374.
Goryshin, I. Y., Jendrisak, J., Hoffman, L. M., Meis, R., and Reznikoff, W. S. (2000). Insertional transposon mutagenesis by electroporation of released Tn5 transposition complexes. *Nature Biotech.* **18,** 97–100.

[4] Identification of Essential Genes in Bacteria

By DAVID R. HILLYARD and MICHAEL J. REDD

Abstract

Essential genes are identified in duplicated regions of the bacterial chromosome. Transposition of a vector that forms operon fusions into a strain carrying a chromosomal duplication allows insertion of the transposon into essential genes because a second copy of the essential gene is present. When the duplication is allowed to segregate, only the segregant that carries the copy of the intact essential gene survives. The transposon insertion in the essential gene is maintained only in the duplication derivatives. A technique is described that uses a Tn*10* derivative, Tn*10dTc-araC*$^+$, which contains a cloned copy of the *Escherichia coli araC*$^+$ gene, as a portable region of homology to generate large duplications of the *Salmonella* chromosome. The duplication is maintained in the population by growth in the presence of tetracycline. When the *lac* operon fusion vector, MudJ, is transposed into the duplicated region, removal of tetracycline from the growth media allows segregation of the duplication yielding (Ara$^-$) haploid segregants which appear as red colonies or as red/white (Ara$^{-/+}$) sectoring colonies on TTC arabinose indicator plates. However, if the insertion is in an essential gene, only segregants that lose the MudJ insertion in the essential gene survive. In this case, selection for the insertion in the essential gene yields solid white (Ara$^+$) colonies in the absence of tetracycline. While the specific design presented uses Mud transposon insertions to generate *lac* operon (transcriptional) and *lacZ* gene (translational) fusions to essential genes, this technique can be used to generate transposon insertions of any kind into essential genes.

Introduction

The arrival of antibiotic-resistance strains of pathogenic bacteria has forced a major push for the development of novel drugs to combat these resistant strains. Identification of essential genes in any bacterial species is an important step for the identification of targets for new generations of

METHODS IN ENZYMOLOGY, VOL. 421   0076-6879/07 $35.00
                     DOI: 10.1016/S0076-6879(06)21004-8

antibiotic drugs. Several techniques such as signature-tagged mutagenesis (Chiang *et al.*, 1999; Hensel *et al.*, 1995) and genetic footprinting (Smith *et al.*, 1995, 1996) have provided mechanisms for the identification of essential genes for any growth condition. Both techniques employ transposon-based strategies. Genome-wide transposon mutagenesis is also used to identify essential genes by the failure to isolate transposon insertions in a given gene (Jacobs *et al.*, 2003). Presented here is a method for the identification of transposon insertions in bacterial genes encoding essential functions. An advantage that this method has over other systems is the generation of reporter fusions for studying the regulation of the essential genes. This chapter will focus on the *lac* operon fusion vector, MudJ, but any transposon that confers a selection will work. The method reported here relies on the creation of merodiploid strains in the form of tandem chromosomal duplications. Insertions into essential genes within the duplicated chromosomal segment are viable and are identified in a simple sectoring assay as colonies that are unable to generate haploid segregants in the presence of antibiotic selection.

Tandem duplications arise spontaneously in bacterial and bacteriophage populations and can be present in a few percent of a growing population of cells (Anderson and Roth, 1977, 1981). Tandem chromosomal duplications with defined genetic endpoints can also be generated in the laboratory as discussed in Chapter 7. Transposons provide portable regions of homology for recombination events that unite different regions of the chromosome in ways that generate tandem duplications with the transposon left inserted at the joint point of the duplication. The duplication is maintained by selection for the transposon-encoded antibiotic resistance, and removal of selection allows for recombination between the duplicated segments yielding haploid segregants that lose the transposon held at the join point. The duplication can also be maintained by loss of homologous recombination capability through disruption of the *recA* gene. Figure 1 illustrates the generation of a tandem duplication that encompasses the regions of the chromosome between points adjacent to the *his* and *nadB* operons. Unlinked insertions of a Tn*10* derivative, Tn*10-araC*$^+$, provide the regions of homology to generate the duplication via RecA-mediated recombination. The sites of the original Tn*10-araC*$^+$ insertion mutants define the duplication endpoints. A Tn*10* insertion linked to any auxotrophic marker (such as the *nadB*::Mud insertion in Fig. 1) can be used to generate a duplication through recombination with any homologous Tn*10* insertion located on the opposite side of the selected gene. In Fig. 1, the Tn*10-araC*$^+$ inserted clockwise of *nadB* (58 min on the *Salmonella* linkage map [Sanderson *et al.*, 1995]) will generate duplications with Tn*10-araC*$^+$ insertions located counterclockwise to

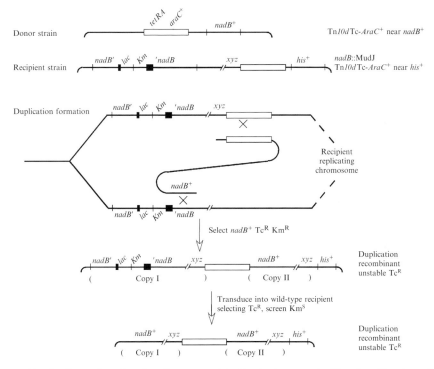

FIG. 1. Generation of tandem chromosomal duplication between the *his* and *nadB* regions of the *S. enterica* chromosome using transposon Tn*10d*Tc-AraC⁺ as a portable region of homology. The donor strain carries Tn*10d*Tc-AraC⁺ insertion near *nadB*⁺, but on the chromosomal side of *nadB*⁺ that is opposite to the *his*⁺ locus. The recipient is *araC*⁻ at the chromosomal *araC* locus, which carries an antibiotic-resistant insertion in the *nadB* gene (a *nadB*::MudJ is used for an example) and has a Tn*10d*Tc-AraC⁺ insertion near *his*⁺. Transduction of the recipient to nadB⁺ occurs in the presence of Tc. The nadB⁺ transductants that remain Tc^R and Km^R result in duplication formation of the region between the *nadB* and *his* regions of the chromosome as diagrammed with a recombinant Tn*10d*Tc-AraC⁺ insertion at the duplication join point. Transduction of the Tn*10d*Tc-AraC⁺ insertion into a *araC*⁻ recipient results in a strain that carries the Tn*10d*Tc-AraC⁺ held tandem duplication without the *nadB* insertion allele (*nadB*::MudJ).

nadB, such as the Tn*10-araC*⁺ insertion near *his* (45 min). Conversely, a Tn*10-araC*⁺ counterclockwise of *nadB* will generate duplications with Tn*10-araC*⁺ insertions located clockwise to *nadB* on the chromosome. The selection also requires that the Tn*10* elements be inserted in the same orientation in the chromosome. Use of the Tn*10-araC*⁺ provides a screen for segregation of the duplication when selection is removed by growing the duplicated strain in the absence of Tc. Large duplications (as illustrated if Fig. 1) segregate at a frequency that is high enough to detect duplication segregation within each colony on indicator medium. This is because about 13% of the genome is duplicated and thus available for homologous recombination to yield haploid segregants.

General Considerations

Construction of Tandem Duplications

Duplication formation requires the presence of homologous DNA elements inserted at unique positions and in the same orientation on the chromosome. The tandem duplications generated in this way have the same gene sequence orientation within the chromosome joined by the hybrid homologous DNA element used in the cross. Insertions of Tn*10-araC*$^+$ that flank a region to be duplicated are used as the portable regions of homology. First, Tn*10-araC*$^+$ insertions from random insertion pools are mapped relative to selectable auxotrophic markers near the desired duplication endpoints. The co-transduction frequency between Tn*10* and the gene of interest can be used to estimate the distance between the insertion and the gene of interest (Sanderson and Roth, 1988). Primers can then be designed based on linkage data for PCR mapping of the Tn*10-araC*$^+$ clockwise or counterclockwise of the auxotrophic gene and the orientation of the Tn*10-araC*$^+$ element in the chromosome. Duplications generated between Tn*10-araC*$^+$ elements require that the elements be inserted in the same orientation in the chromosome.

Once Tn*10-araC*$^+$ insertions have been identified that flank the gene of interest and are in the same orientation on the chromosome, they can be used to generate deletions and duplications between the points of insertion as diagrammed in Fig. 1. The selection for duplications requires one endpoint to be marked with a Tn*10-araC*$^+$ insertion linked by co-transduction to a selectable marker in the donor (*nadB*$^+$ in Fig. 1). The recipient requires the second endpoint to be marked with a Tn*10-araC*$^+$ on the opposite side of the selectable marker relative to the donor Tn*10-araC*$^+$ insertion. The example in Fig. 1 has the donor's Tn*10-araC*$^+$ element inserted clockwise of the *nadB* locus, and the recipient's Tn*10-araC*$^+$ element inserted clockwise of the *nadB* locus. The selected marker in the recipient is marked with a drug-resistant null allele (*nadB*::Mud). Then, selection for repair of the auxotrophic marker (*nadB*$^+$) in the recipient while holding and Km^R of the *nadB*::MudJ can only occur by the recombination diagrammed in Fig. 1 yielding a duplication between the points of insertion of the Tn*10-araC*$^+$ elements from the donor and recipient strains. The hybrid Tn*10-araC*$^+$ at the duplication join point illustrated in Fig. 1

connects regions of the chromosome that are normally separated by more than 6000 kbp of DNA.

Construction of Tn10dTc-araC⁺

Plasmid pNK217 carries the smallest Tc^R derivative of Tn*10*, Tn*10 del-16 del-17* (Tn*10d*Tc), lacking transposase that can be complemented for transposition by providing transposase (Foster *et al.*, 1981). We cloned the *araC⁺* gene from *E. coli* into the *HinDIII* site of Tn*10d*Tc to create Tn*10d*Tc-*araC⁺* (plasmid pDH10). Strains defective in the *araC* gene are complemented by Tn*10d*Tc-*araC⁺*. This provides a simple screen on arabinose-utilization plates for the presence or absence of the Tn*10d*Tc-*araC⁺* derivative.

Plan of the Experiment

Isolation of Mud Insertions in and Tn10-araC⁺ Near Genes of Interest

Mud: The random mutagenesis of *Salmonella* by MudJ transposition is described in Chapter 7. MudJ transposition is selected by selecting for MudJ-encoded Km^R. When the selection for MudJ insertions is done on rich medium, about 4% of the insertions are in auxotrophic genes. The biosynthetic pathway ascribed to a specific auxotrophic MudJ insertion is determined by a technique called auxonography (Davis *et al.*, 1980). For some pathways such as the *his* biosynthetic genes, all the genes are in a single nine-gene operon; consequently, the chromosomal location of a *his*::MudJ insertion is known immediately. For biosynthetic pathways comprised of unlinked genes, linkage to known markers near specific genes is usually required. Also, PCR with primers designed to different loci can be used to determine the gene location of different auxotrophic MudJ insertions. Insertions defective in carbon or nitrogen source utilization must be screened for according to the specific carbon or nitrogen source.

Tn*10d*Tc-*araC⁺*: Plasmid pDH10 was transformed into *Salmonella* strain SL4213 (a restriction-deficient *Salmonella* strain). P22-transducing phage was then grown on pDH10/SL4213 and the resulting lysate used to transduce a recipient strain expressing the Tn*10* transposase from plasmid pNK972 (Way *et al.*, 1984). Transposition recombinants were selected for by selecting for Tn*10d*Tc-*araC⁺*-encoded Tc^R. Tens of thousands of Tc^R transposition recombinant colonies were pooled, and a P22-transducing lysate prepared on the randomized pool of insertion mutants. Individual Tn*10d*Tc-*araC⁺* insertions were identified by linkage to known auxotrophic Mud insertion mutants throughout the chromosome.

Construction of Tn10dTc-araC+ Join-Point Duplications

The construction of a Tn*10*dTc-*araC*+ joint-point duplication spanning the interval from approximately 45′ to 58′ on the *Salmonella* chromosome (*his* to *nadB*) is illustrated in Fig. 1. A recipient *Salmonella* chromosome contains a *nadB*::MudJ insertion (Lac+ operon fusion [blue in the presence of Xgal]), as well as a Tn*10*dTc-*araC*+ insertion linked to the *nadB* side of the *his* operon (clockwise on the standard linkage map). A P22-transducing lysate was prepared on a donor strain containing a Tn*10*dTc-*araC*+ insertion linked to and counterclockwise of *nadB*+. Both Tn*10*dTc-*araC*+ insertions (donor and recipient) are in the same orientation relative to the bacterial chromosome. Selection for *nadB*+ prototrophy results in two classes of recombinants– simple replacement of *nadB*::MudJ, and tandem duplication formation in which one copy of the *nadB*::MudJ insertion remains in the recipient chromosome. These two classes are distinguished by screening for blue colonies on Xgal-containing plates (10%) or by holding selection for the *nadB*::MudJ insertion (Km^R). The Tn*10*dTc-*araC*+ joint point is subsequently transduced into a *Salmonella araC*-defective strain selecting Tc^R.

Because the joint-point insertion is flanked by sequence from the *his* region on one side and sequences from the *nadB* region on the other side, it can only be inherited by regenerating the original duplication. The duplicated strain is subjected to transposition mutagenesis by MudJ (Hughes and Roth, 1988). Resulting MudJ (Km^R) transposon insertions are screened for those with insertions in essential genes of the duplicated region (the 37′–45′ region of the *Salmonella* chromosome). Because the duplication harbors two tandem copies of this large chromosomal segment, cells grown in the absence of tetracycline lose the duplication at a high frequency, and are visualized as sectored colonies on arabinose-indicator plates because segregation events result in the loss of the Tn*10*dTc-*araC*+ element at the joint point. When the duplicated strain is plated in triphenyl tetrazolium chloride-arabinose (TTC-Ara) medium, the resulting colonies produce white (Ara+) and red (Ara−) sectors as well as red-only (Ara−) colonies. Growth on TTC-Ara plates with added Tc results in nonsectoring white (Ara+) colonies because the segregated cells are Tc^S Ara− and do not grow.

Identification of Essential Genes and Isolation of lac Fusions to Promoters of Essential Genes of Salmonella

The strategy for identification of MudJ (or MudK) transposon insertions in essential *Salmonella* genes is illustrated in Fig. 2. A duplication-containing strain held by a Tn*10*dTc-*araC*+ insertion at the duplication joint point is mutagenized with transposon MudJ. On TTC-Ara plates

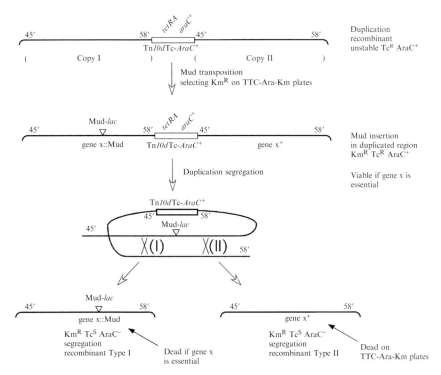

FIG. 2. Isolation of Mud insertions in essential genes. A Tn*10d*Tc-AraC$^+$ held tandem chromosomal duplication strain is subject to MudJ or MudK transposition. If the Mud inserts into an essential gene within the tandem duplicated segment, these cells will not segregate haploid segregants containing the Mud insertion mutation. If the Mud element is held by growing in the presence of Km, then the cells must maintain the second duplicated segment that carries the wild-type essential gene and do not show visible segregation (the Km^R segregants are not viable) on TTC-Ara-Km plates (AraC$^+$-only colonies). If the Mud element is inserted into a nonessential gene, Km^R segregants are viable, and these cells show visible segregation (the Km^R segregants are viable) on TTC-Ara-Km plates AraC$^-$ and AraC$^{-/+}$.

(No Tc), the duplication segregates at high frequency and each colony is Ara$^-$ (red) or has Ara$^-$ (red) sectors. There are no solid white (Ara$^+$) colonies. The duplicated strain is grown in nutrient broth containing tetracycline to maintain selection for the retention of duplicated chromosomal segments. This strain serves as a recipient for MudJ mutagenesis by P22 transduction. P22 grown on a MudJ hopper strain is used to transduce the duplication recipient to MudJ-encoded Km^R (Hughes and Roth, 1988).

The MudJ transposon confers kanamycin resistance, and in one orientation leads to the formation of *lac*-operon fusions.

Three classes of viable insertions are anticipated: (1) insertions in non-essential genes in nonduplicated chromosomal regions, and insertions in (2) nonessential or (3) essential genes in either copy of the duplicated chromosomal segment. The latter class is identified as bacteria that grow as white, nonsectoring colonies on TTC arabinose plates containing Km and lacking Tc. If an MudJ insertion is obtained in a gene within the duplicated segment, a second wild-type copy of the gene is present in the second duplicated segment. Haploid segregants are obtained when the strain is grown in the absence of tetracycline that either retain (Fig. 2, recombination event I) or lose the MudJ insertion allele (recombination event II).

For insertions in essential genes, only haploid segregants resulting from recombination event I (no MudJ insertion) are viable. Plating on TTC-Ara plates without Tc to hold the duplication, yields red (Ara$^-$) colonies or white colonies with red sectors. Plating on TTC-Ara-Km plates selects for colonies that retain the MudJ insertion.

For MudJ insertions in essential genes within the duplicated segment, red or sectored colonies do not appear and the colonies are white-only (Ara$^+$) because selection for the MudJ insertion in the essential gene (Km^R) demands that the duplication (Ara$^+$) is maintained.

Procedures

Procedure 1. Growing P22 Lysates

Materials

LB (Luria broth): Per liter deionized water, 10 g tryptone, 5 g yeast extract, 5 g NaCl.

Ex50 salts: 50% D-glucose.

Sterile saline: Per liter deionized water, 8.5 g NaCl.

Top agar: Per liter deionized water, 10 g tryptone, 7 g agar.

P22 broth: 200 ml LB, 2 ml Ex50 salts, 0.8 ml 50% D-glucose, 10^7 to 10^8 plaque-forming units (pfu)/ml P22 transducing phage: P22 HT/*int*.

P22 HT/int lysate preparation. Grow a P22-sensitive host strain to saturation in LB. Make serial dilutions of a P22 lysate in sterile saline and plate 0.1 ml of diluted phage with 0.1 ml of cell culture in 3 ml top agar on a LB agar plate (12 g of agar per liter). Pick a single plaque with a pasteur pipette and inoculate a 1-ml LB-saturated culture of a sensitive strain. Add 4 ml of P22 broth that does not have added P22, and grow with shaking at 37° for 5 or more hours (lysates left over the weekend will work, but usually an all-day or overnight incubation period is used for convenience). Titer the resulting lysate and use it to prepare a working stock of P22 broth. For all future lysates, add 4 ml of P22 broth to 1 ml LB saturated culture of a sensitive strain and grow 5 to 36 hours at 37° with shaking. Pellet cells by centrifugation (10 min at full speed in a table-top centrifuge, or for larger volumes spin 5 min at 8000 rpm in a SS34 rotor). Decant the supernatant into a sterile tube, add $CHCl_3$, and vortex to sterilize. Store at 4°.

Procedure 2. Preparation of a P22-Transducing Lysate Prepared on 50,000 Random Tn10dTc-araC+ Insertion Mutants

Materials

Media: LB (Luria broth)–Per liter deionized water, 10 g tryptone, 5 g yeast extract, 5 g NaCl; add 15 g/l Bacto agar for solid medium.

Antibiotics (final concentration): sodium-ampicillin (100 μg/ml for selection for pDH10 plasmid), tetracycline-HCl (15 μg/ml in LB).

Donor: P22 HT/int–transducing lysates on Salmonella enterica strain with plasmid pDH10 (Tn10dTc-araC+).

Recipients: Wild-type S. enterica strain LT2 carrying the Tn10 transposase-expressing plasmid pNK972 (Way et al., 1984).

1. Start 1-ml overnight culture of the S. enterica strain pNK972/LT2 in LB plus Ampicillin (Ap). Grow overnight with aeration at 37°.

2. First do a test cross to determine the number of Tn10dTc-araC+ insertions obtained per 0.1 ml of diluted phage stock. Dilute the Tn10dTc-araC+ donor lysate 10^{-1}, 10^{-2}, and 10^{-3} into LB. In sterile tubes, mix 0.1 ml of cells from the overnight culture with 0.1 ml of phage dilution grown on the Tn10dTc-araC+ donor. Some of the phage particles will inject Tn10dTc-araC+ DNA into your recipient cells. Let sit 1 hour at room temperature for phenotypic expression of Tc^R. Determine a donor phage dilution that will yield ≈5000 colonies per plate.

3. Start fresh 10-ml overnight cultures of pNK972/LT2 in LB plus Ap. Grow overnight with aeration at 37°. In separate sterile tubes, add 1.2 ml of the appropriate phage dilution to 1.2 ml of cells. Let sit 1 hour at room temperature. Plate 0.2 ml of the mixture onto each of 10 L-Tc plates. Incubate overnight at 37°. Add 1 ml of LB to each plate, and with a sterile glass rod mix the colonies to pool Tc^R colonies resulting from Tn10dTc-araC+ transposition into the recipient chromosome. Transfer cells from each plate into a single sterile tube to pool the cells from all 10 plates creating a pool of ≈50,000 independent Tn10dTc-araC+ transposon-insertion mutants. Wash the cell pool by pelleting in an SS34 rotor (8000 rpm for 5 min), and resuspending the

cell pellet in fresh LB. Dilute the pooled cells into 20 ml LB to $\approx 2 \times 10^9$ cells/ml (the density of a fresh overnight culture). Add 80 ml of P22 broth and grow 6 to 24 hours at 39° with aeration. Pellet debris (8000 rpm in SS34 rotor for 5 min), and transfer the supernatant (P22-transducing lysate) to a sterile bottle. Add $CHCl_3$ and shake vigorously to sterilize. Store at 4°. This is a P22-transducing lysate grown on a pool of 50,000 random Tn10dTc-araC$^+$ insertion mutants inserted throughout the S. enterica chromosome.

Procedure 3. Isolation of Tn10dTc-araC$^+$ Insertion Mutants at Duplication Endpoints Near his and nadB

Materials

Media: Minimal E salts (Davis et al., 1980) with 0.2% D-glucose and 15 g/l Bacto agar for solid media.

LB (Luria broth): Per liter deionized water, 10 g tryptone, 5 g yeast extract, 5 g NaCl; add 15 g/l Bacto agar for solid medium.

TTC-indicator plates: Per liter deionized water, 10 g tryptone, 1 g yeast extract, 5 g NaCl, 15 g Bacto agar, 1% sugar, and 50 mg TTC. To prepare TTC-*ara* indicator plates, add 10 g tryptone, 1 g yeast extract, 5 g NaCl, and 15 g Bacto agar to 500 ml of deionized water in a 1-liter flask (Flask A). Add 10 g of L-arabinose to 500 ml of deionized water in a second 1-liter flask (Flask B). Sterilize by autoclave and cool to $\approx 55°$; add 5 ml of a 1% solution of TTC (2,3,5 triphenyl tetrazolium chloride). Mix and pour. The chromogenic substrate for β-galactosidase activity, Xgal, was added to 25 μg/ml in minimal E medium and 40 μg/ml in LB medium.

Antibiotics (final concentration): Kanamycin-SO$_4$ (50 μg/ml in LB or TTC medium, 100 μg/ml in minimal E glucose medium), sodium-ampicillin (100 μg/ml for selection for pDH10 plasmid), tetracycline-HCl (15 μg/ml in LB or TTC-Ara and 7.5 μg/ml in minimal E medium).

Donor: P22 HT/*int* transducing lysate on S. enterica pool of 50,000 independent Tn10dTc-araC$^+$ insertion mutants (Procedure 2).

Recipients: S. enterica recipient strains defective in either nadB (nicotinic acid growth requirement) or his (histidine growth requirement).

1. Do a test cross to determine the number of Tn10dTc-araC$^+$ insertions obtained per 0.1 ml of diluted Tn10dTc-araC$^+$ insertion pool lysate. Start 1-ml cultures of the S. enterica nadB and his auxotrophs in LB.

2. Dilute the Tn10dTc-araC$^+$ pool lysate 10^{-2}, 10^{-3}, and 10^{-4} into LB. Add 0.1 ml of recipient cells followed by 0.1 ml of phage lysate dilution directly onto L-Tc plates. Prepare one plate for each dilution.

Also, prepare cell-only and phage-only control plates. Determine a donor phage dilution that will yield 1000 colonies per plate.

3. Start fresh 1-ml overnight cultures of the *S. enterica nadB* and *his* auxotrophs in LB. Grow overnight with aeration at 37°. In separate sterile tubes, add 0.6 ml of the appropriate phage dilution to 0.6 ml of cells. Plate 0.2 ml of the mixture onto each of 5 L-Tc plates. Incubate overnight at 37°.

4. Replica print to E-Tc plates (minimal E glucose), and incubate overnight at 37°. Expect about 0.5% of the Tc^R recombinants to be linked to either auxotrophic marker, and transduce the recipients to proto-trophy. These will grow on the E-Tc plates.

5. For each recipient, pick 10 prototrophic (NadB$^+$ or His$^+$ Tc^R) transductants, and make phage-free on either green or EBU indicator plates (Maloy *et al.*, 1996).

6. Prepare P22-transducing lysates on each insertion mutant. Add 4 ml P22 broth to 1 ml of an overnight culture of cells, grow more than 5 h at 37° with aeration, pellet cells, transfer supernatant (phage lysate) to sterile tube, add ChCl$_3$, vortex, and store at 4°.

7. Determine linkage (percent co-transduction) of each Tn*10*-*d*Tc- *araC*$^+$ insertion to either *nadB*$^+$ or *his*$^+$. Grow fresh 2-ml overnight cultures of the *S. enterica nadB* and *his* auxotrophs in LB. Dilute each lysate 10^{-3} in LB. Add 0.1 ml cells to an L-Tc plate followed by 0.1 ml of diluted phage donor. Use the *nadB* recipient for Tn*10d*Tc-*araC*$^+$ donor insertions linked to *nadB*$^+$, and use the *his* recipient for Tn*10d*Tc-*araC*$^+$ donor insertions linked to *his*$^+$. Incubate overnight at 37°.

8. For each transduction, pick and transfer (patch) 100 Tc^R trans-ductants onto L-Tc plates. Incubate overnight at 37°.

9. Replica print to E-Tc and L-Tc plates. Incubate overnight at 37°.

10. The number of patches that grow on E-Tc plates is the percent co-transduction between the individual Tn*10d*Tc-*araC*$^+$ insertion mutant and either *nadB*$^+$ or *his*$^+$.

11. Based on the percent co-transduction, determine the physical dis-tance between each Tn*10d*Tc-*araC*$^+$ insertion mutant and either *nadB*$^+$ or *his*$^+$ (Sanderson and Roth, 1988).

12. Based on the physical distances, design primers to genes on either side of the *nadB* and *his* loci to PCR; amplify using DNA from each Tn*10d*Tc-*araC*$^+$ insertion mutant to determine exact positions in the chromosome.

13. Determine chromosomal orientation of each Tn*10d*Tc-*araC*$^+$ in-*tetRA* or *araC* sequence within the Tn*10d*Tc-*araC*$^+$ element and flank-ing chromosomal DNA.

Procedure 4. Generation of Tandem Chromosomal Duplications Between his *and* nadB *Regions of Chromosome with Tn10dTc-araC⁺ Insertion at Duplication Join Point*

Materials

P22 HT/*int*–transducing lysates on *S. enterica* strains carrying a Tn*10-dTc-araC⁺* insertion near the *his⁺* locus, and one near and counter-clockwise of *nadB⁺* (Procedure 3).

S. enterica strains (1) *araC⁻* and (2) *araC⁻ nadB499*::MudJ.

1. Start a 1-ml overnight culture of strain *S. enterica araC⁻ nadB499*::MudJ. Grow overnight with aeration at 37°.

2. Transduce a Tn*10dTc-araC⁺* insertion linked to the *his* operon from Procedure 3 into the *S. enterica araC⁻ nadB499*::MudJ strain selecting for Tc^R on L-Tc plates. Incubate overnight with aeration at 37°.

3. Purify a phage-free Tc^R transductant on green or EBU indicator plates (Maloy *et al.*, 1996). Check to be sure that strain remains Km^R. This will be the recipient strain for the duplication construction (*nadB*:: MudJ Tn*10dTc-araC⁺* near *his⁺*) (Fig. 1).

4. Using the P22 lysate prepared on the strain carrying a Tn*10d* Tc-*araC⁺* insertion near the *his⁺* locus in the same chromosomal orienta-tion as the Tn*10dTc-araC⁺* insertion in the recipient strain, transduce the recipient strain to *nadB⁺* on E-glucose–Tc-Xgal plates.

5. Purify phage-free Tc^R transductants on green-Tc or EBU-Tc indicator plates (Maloy *et al.*, 1996).

6. Screen for presence of the duplication by streaking for single colonies on TTC-Ara and TTC-Ara-Tc plates. Strains carrying tandem duplications will segregate Red (Ara⁻) colonies on TTC-Ara plates while remaining white (Ara⁺) on TTC-Ara-Tc plates. The red colonies should also be Tc^S due to loss of the Tn*10dTc-araC⁺* insertion at the duplication join point (see Fig. 2).

7. Prepare a P22-transducing lysate on the duplication construct. Grow 1-ml of cells overnight in LB-Tc, add 4 ml P22 broth, and harvest lysate after at least 5 h growth at 37°, as described in above descriptions of P22-lysate preparations. Store with CHCl₃ at 4°.

8. Using P22 lysate prepared on the duplication construct, transduce the *S. enterica araC⁻* strain to Tc-resistance on L-Tc; isolate a phage-free Tc^R transductant. Screen for Km^S. (Because the Tn*10dTc-araC⁺* inser-tion at the duplication join point is not linked to the *nadB*::MudJ insertion, only the Tn*10dTc-araC⁺* insertion will be inherited.) In addition, screen for segregation of red or sectored colonies on TTC-Ara plates. The final strain will serve as the tandem duplication recipient for the isolation of

MudJ (*lac* transcriptional fusion vector) or MudK (*lacZ* translational fusion vector) insertions in essential genes.

Procedure 5. Isolation of Mud Insertions in Essential Genes Between his *and* nadB *Regions of Chromosome*

Materials

P22 HT/*int* transducing lysates on MudJ (TT10288) or MudK (TT10381) hopper strains (Hughes and Roth, 1988).

S. enterica strain carrying a tandem duplication of the *his* to *nadB* region of the chromosome (from Procedure 4).

1. Perform a test cross to determine the number of Mud insertions obtained per 0.1 ml of diluted phage stock. Start a 1-ml overnight culture of *his-nadB* duplication strain in L-Tc. Grow overnight with aeration at 37°. Dilute the Mud donor lysates 10^{-1}, 10^{-2}, and 10^{-3} into LB. In sterile tubes, mix 0.1 ml of cells from the overnight culture with 0.1 ml of phage dilution grown on either the MudJ or MudK donor. Let sit 1 hour at room temperature for phenotypic expression of MudJ/K-encoded Km^R. Plate on TTC-Ara-Km plates. Determine a donor phage dilution that will yield ≈500 to 1000 colonies per plate.

2. Start a 25-ml overnight culture of *his-nadB* duplication strain in L-Tc. Grow overnight with aeration at 37°. Add 11 ml of diluted MudJ donor lysate to 11 ml of cells. Separately, add 11 ml of diluted MudK donor lysate to 11 ml of cells. Let sit 1 hour at room temperature for phenotypic expression of MudJ/K-encoded Km^R. For each transposition experiment, plate 0.2 ml onto 100 TTC-Ara-Km plates. Incubate overnight at 37°.

3. Pick any nonsectoring Ara^+ (white) Km^R colonies, and make phage-free on green or EBU plates with added Km (50 μg/ml) to maintain the duplication.

4. Screen the putative Mud insertion mutants in essential genes for *lac* transcriptional or translational fusions to the promoter or gene to which the Mud has inserted on lactose indicator medium (LB-Xgal, MacConkey lactose, or TTC-lactose plates). Whether a fusion is Lac^+ on a given indicator depends on the level of β-gal activity. It takes more β-gal activity to be Lac^+ on TTC-lactose than on MacConkey-lactose than on LB-Xgal, and because it takes very little β-gal activity to be blue on Xgal, this can lead to false positives.

5. DNA sequence analysis using semirandom PCR with primers to the ends of the MudJ or MudK elements is used to determined the gene to which the Mud is inserted, and if it is in the correct orientation (and

reading frame) to make a *lac* operon (or *lacZ* gene) fusion to the essential gene.

References

Anderson, P., and Roth, J. (1981). Spontaneous tandem genetic duplications in *Salmonella typhimurium* arise by unequal recombination between rRNA (rrn) cistrons. *Proc. Natl. Acad. Sci. USA* **78**, 3113–3117.

Anderson, R. P., and Roth, J. R. (1977). Tandem genetic duplication in phage and bacteria. *Annu. Rev. Microbiol.* **34**, 473–505.

Chiang, S. L., Mekalanos, J. J., and Holden, D. W. (1999). *In vivo* genetic analysis of bacterial virulence. *Annu. Rev. Microbiol.* **53**, 129–154.

Davis, R. W., Botstein, D., and Roth, J. R. (1980). "Advanced Bacterial Genetics." Cold Spring Harbor, NY: Cold Spring Harbor Laboratory.

Foster, T. J., Davis, M. A., Roberts, D. E., Takeshita, K., and Kleckner, N. (1981). Genetic organization of transposon Tn10. *Cell* **23**, 201–213.

Hensel, M., Shea, J. E., Gleeson, C., Jones, M. D., Dalton, E., and Holden, D. W. (1995). Simultaneous identification of bacterial virulence genes by negative selection. *Science* **269**, 400–403.

Hughes, K. T., and Roth, J. R. (1988). Transitory cis complementation: A method for providing transposition functions to defective transposons. *Genetics* **119**, 9–12.

Jacobs, M. A., Alwood, A., Thaipisuttikul, I., Spencer, D., Haugen, E., Ernst, S., Will, O., Kaul, R., Raymond, C., Levy, R., Chun-Rong, L., Guenthner, D., Bovee, D., Olson, M. V., and Manoil, C. (2003). Comprehensive transposon mutant library of *Pseudomonas aeruginosa*. *Proc. Natl. Acad. Sci. USA* **100**, 14339–14344.

Maloy, S. R., Stewart, V. J., and Taylor, R. K. (1996). "Genetic Analysis of Pathogenic Bacteria." Cold Spring Harbor, NY: Cold Spring Harbor Press.

Sanderson, K. E., and Roth, J. R. (1988). Linkage map of *Salmonella typhimurium*, edition VII. *Microbiol. Rev.* **52**, 485–532.

Sanderson, K. E., Hessel, A., and Rudd, K. E. (1995). Genetic map of *Salmonella typhimurium*, edition VIII. *Microbiol. Rev.* **59**, 241–303.

Smith, V., Botstein, D., and Brown, P. O. (1995). Genetic footprinting: A genomic strategy for determining a gene's function given its sequence. *Proc. Natl. Acad. Sci. USA* **92**, 6479–6483.

Smith, V., Chou, K. N., Lashkari, D., Botstein, D., and Brown, P. O. (1996). Functional analysis of the genes of yeast chromosome V by genetic footprinting. *Science* **274**, 2069–2074.

Way, J. C., Davis, M. A., Morisato, D., Roberts, D. E., and Kleckner, N. (1984). New Tn10 derivatives for transposon mutagenesis and for construction of lacZ operon fusions by transposition. *Gene* **32**, 369–379.

[5]   Isolation of Transposon Insertions

By STANLEY R. MALOY

Abstract

Transposon insertions in and near a gene of interest facilitate the genetic characterization of a gene *in vivo*. This chapter is dedicated to describing the isolation of mini-Tn10 insertions in any desired nonessential gene in *Salmonella enterica*, as well as the isolation of mini-Tn10 insertions near particular genes. The protocols describe use of a tetracycline-resistant Tn10 derivative, but similar approaches can be used for derivatives resistant to other antibiotics. In addition, these approaches are directly applicable to other bacteria that have generalized transducing phages.

Introduction

Most transposons have low target-site specificity, allowing insertions in many sites throughout the bacterial genome. Antibiotic-resistant transposons are particularly useful genetic tools because the antibiotic resistance provides a selectable marker within the transposon. This chapter will focus on antibiotic-resistant transposons: transposons derived from Tn*10*. A few of the applications of antibiotic resistant transposon insertions are summarized in Chapter 2.

Transposition typically occurs at a low frequency *in vivo*. Hence, an effective delivery system is required to isolate a representative collection of transposon insertions in a bacterial genome. An effective delivery system demands (1) a high frequency of transfer into recipient cells, (2) a strong selection for the transposon insertion with a minimal background of spontaneous mutants that answer the selection applied, (3) removal of donor DNA within the recipient after the transposition event, and (4) a mechanism to prevent further unwanted transposition events once the desired insertion is isolated.

Experimental Rationale

The first protocol below describes the isolation of transposon Tn10dTet pools in *S. enterica* by delivery of a relaxed-specificity transposase from a multiple-copy number plasmid in trans. Once the transposon pools are isolated, a phage lysate is grown on these strains to segregate the transposon insertion from the transposase. The phage lysate grown on transposon

METHODS IN ENZYMOLOGY, VOL. 421
0076-6879/07 $35.00
DOI: 10.1016/S0076-6879(06)21005-X

pools can then be used to transfer the transposon insertions into a new strain background by selection for the transposon-associated antibiotic resistance, in this case tetracycline resistance. Assuming a random distribution of insertion of the transposon throughout the genome, roughly 1 in 2000 of the mutants will have an insertion in any given nonessential gene. The desired mutants can thus be obtained by selection for a particular phenotype, or by screening through several thousand independent tetracycline-resistant insertions.

The second protocol below describes the isolation of Tn10dTet insertions linked to an auxotrophic mutation. It is often useful to have an antibiotic-resistant insertion near, but not in, a particular gene. The transposon pools can also make selection for such linked insertions possible. A transducing lysate grown on the transposon pool is used to transduce a strain with an observable phenotype near the desired insertion site (e.g., an auxotrophic mutation), selecting for the transposon-associated antibiotic resistance, and screening for repair of the adjacent phenotype. Assuming a random distribution of insertion of the transposon throughout the genome and a transducing phage that can package approximately 1% of the genome per phage head, roughly 1 in 200 of the mutants will have an insertion closely linked to a particular genetic locus.

Protocol 1. Isolation of Mini-Tn10 (Tn10dTet) Insertion Pools

1. Start a single colony of the *S. enterica* donor strain carrying pNK976 with *ats* transposase in LB broth plus ampicillin. Incubate overnight at 37°.
2. Prepare a phage P22 *int* lysate on the donor strain (Maloy, 1990).
3. Start a single colony of the recipient strain *S. enterica* LT2 (wild-type) in LB broth. Incubate overnight at 37°.
4. Add the recipient cells and the P22 lysate in sterile microfuge tubes as shown in Table I.
5. Gently mix each tube, and then incubate at 37° for about 20 min to allow phage adsorption.

TABLE I
PROTOCOL 1: P22 DELIVERY OF TRANSPOSON

Plate	Recipient bacteria	P22 HT *int* donor	Plate descriptor
A	500 μl	—	No phage control
B	—	100 μl	No bacteria control
C	500 μl	100 μl	Pool 1
D	500 μl	100 μl	Pool 2
E	500 μl	100 μl	Pool 3
F	500 μl	100 μl	Pool 4
G	500 μl	100 μl	Pool 5

6. Add 2 ml of LB plus 10-mM EGTA to prevent further phage infection, and gently mix. Spread 0.2 ml from the no-phage control and no-cell control tubes onto separate LB plus tetracycline plates. Spread 0.2 ml from tubes C, D, and E onto separate LB-plus-tetracycline plates. Incubate overnight at 37°.

7. No colonies should have grown on the control plates. (Growth on control plates implies contamination, so the experiment should be scrapped and the problematic culture restarted.) The pool plates should each contain approximately 200 colonies per plate.

8. Add 1 ml of sterile 0.85% NaCl to the surface of each plate. Using a sterile spreader, resuspend the colonies into the liquid. The cell suspension should be dense.

9. Using a sterile pipetor, collect the cell suspension and place the cells from each of the five pools into a separate tube.

10. Centrifuge the cell suspension to collect the cell pellet.

11. Remove the supernatant and resuspend the pellet from each separate pool in 20 ml of 0.85% NaCl. Re-centrifuge to pellet the cells.

12. Resuspend each pellet in 1 ml of LB broth. Each cell suspension represents a separate pool of random Tn10dTet insertions. These pools can be stored frozen in 1-M glycerol at −70°.

13. Grow a phage P22 *int* lysate on each Tn10dTet pool. The resulting P22 transducing phage can be used to select for Tet^R insertion mutants, and then to screen for those mutants with the desired phenotype.

Protocol 2. Isolation of Mini-Tn10 (Tn10dTet) Insertions Linked to an Auxotrophic Mutation

1. Start a single colony of the *S. enterica* recipient strain with a known auxotrophic mutation in LB broth. Incubate overnight at 37°.

2. Spread the recipient cells and the P22 HT *int* lysate grown on Tn10dTet pools on five separate LB-plus-tetracycline plates as shown in Table II.

3. Incubate overnight at 37°.

4. No colonies should have grown on the control plates. (Growth on control plates implies contamination, so the experiment should be scrapped.) There should be about 200 colonies on each of the other plates.

5. Replicate the tetracycline-resistant colonies onto minimal medium. Incubate overnight at 37°.

6. Only transductants that inherited an adjacent wild-type allele that repairs the auxotrophic mutation will grow on the minimal plates. Colonies from each pool may be siblings, but colonies from the independent pools represent distinct insertions. Any transductants

TABLE II
PROTOCOL 2: ISOLATION OF LINKED TRANSPOSONS USING P22 TRANSDUCTION

Plate	Recipient bacteria	P22 HT *int* donor	Plate descriptor
A	100 μl	—	No phage control
B	—	20 μl	No bacteria control
C	100 μl	20 μl	Pool 1 lysate
D	100 μl	20 μl	Pool 2 lysate
E	100 μl	20 μl	Pool 3 lysate
F	100 μl	20 μl	Pool 4 lysate
G	100 μl	20 μl	Pool 5 lysate

that will be saved should be checked to ensure that they are phage-free by streaking on EMU plates and cross-streaking against phage P22 H5 (Maloy *et al.*, 1996).

7. Confirm the linkage and co-transduction frequency by growing phage P22 HT *int* on the tetracycline-resistant prototrophic transductants, and performing a backcross into the original auxotrophic mutant strain.

Testing for Randomness of Transposon Insertions

A simple control to confirm that the collection of transposon insertions includes random mutations at many sites in the bacterial genome is to screen for auxotrophic mutants. Auxotrophic mutations arise from disruption of a large number of different genes encoding biosynthetic precursors, thus provide a large target for mutagenesis (Berlyn *et al.*, 1996; Sanderson *et al.*, 1996). Auxotrophic mutants can be identified by replica plating from the rich medium used to select for antibiotic resistant insertions onto minimal medium. If the auxotrophic mutants are due to insertions at many different sites, the auxotrophs will include mutants with a variety of different auxotrophic phenotypes (resulting in distinct supplements to compensate for the different biosynthetic defects). A collection of auxotrophic mutants can be easily screened for their nutritional requirements by testing for growth on crossed pool plates, a process called auxanography (Davis *et al.*, 1980). Eleven pools contain the most common auxotrophic supplements for major biosynthetic pathways for amino acids, nucleotides, and vitamins. Each supplement is present in two of the eleven pools. A mutant requiring one biosynthetic requirement would grow only on the two pools that contain it. For example, histidine is in pool 1 and pool 7 so a histidine auxotroph will only grow on plates supplemented with pools 1 and 7. Certain auxotrophs require two supplements and thus only grow on the one pool that contains both supplements. For example, purine mutants that require both adenine and thiamine will only grow on plates supplemented with pool 6. Certain auxotrophic requirements are only present in pool 11. It should be possible to identify a variety of distinct auxotrophs by this method.

1. Pick colonies of potential auxotrophs and replica plate or patch onto E medium + glucose + crossed pool plates. Also patch a prototrophic control. Incubate overnight at 37°.
2. Replica plate onto E + glucose plates with each of the "crossed pool" supplements (see the table on the next page). Incubate plates at 37°.
3. Examine the patches and record the growth on each of the crossed pool plates. Determine the auxotrophic requirements using the table below.

The composition of the crossed pool plates is shown in Table III (Davis *et al.*, 1980). The composition of pools 1–5 are listed in the vertical columns and pools 6–10 are listed in horizontal rows. The composition of pool 11 is listed at the bottom. Pool 11 contains a variety of nutrients (mainly vitamins) not included in the other pools. Advice on interpreting the results can be found in the table legend.

TABLE III
Crossed Pool Plates

	1	2	3	4	5	
6	Ade	Gua	Cys	Met	Thi	
7	His	Leu	Ile	Lys	Val	
8	Phe	Tyr	Trp	Thr	Pro	PABA, DHBA
9	Gln	Asn	Ura	Asp	Arg	
10	Thy	Ser	Glu	DAP	Gly	
11	Pyridoxine, Nicotinic acid, Biotin, Pantothenate, Ala					

[a] Standard abbreviations are used for amino acids: DAP, diaminopimelic acid; PABA, p-aminobenzoic acid; DHBA, dihydroxybenzoic acid.

[b] Purine auxotrophs fall into several categories. Some purine mutants only require adenine. Other purine mutants grow on adenosine or guanosine (*purC*, *purE*, or *purH*), allowing growth on pools 1, 2, and 6. Other purine mutants require adenosine + thiamine (*purD*, *purF*, *purG*, or *purI*), restricting growth to pool 6.

[c] Mutants that disrupt *pyrA* require uracil + arginine, and thus only grow on pool 9.

[d] Mutants requiring isoleucine + valine (*ilv*) only grow on pool 7.

[e] Mutants with early blocks in the aromatic pathway will only grow on pool 8, which contains both aromatic amino acids and the the intermediates in aromatic amino acid biosynthesis, p-aminobenzoic acid (PABA) and dihydroxybenzoic acid (DHBA).

[f] Mutants with early blocks in the lysine pathway require lysine + diaminopimelic acid (DAP), and thus only grow on pool 4.

[g] Some *thi* mutants require very small amounts of thiamine. These mutants often acquire enough thiamine from pools 5 and 6 to grow on all subsequent pool plates.

[h] Some mutants require either cysteine or methionine, and thus grow on pools 3, 4, or 6.

[i] There is insufficient glutamine in LB to supplement mutants that require high concentrations of glutamine (e.g. *glnA* mutants).

[j] TCA cycle mutants may have complex requirements that cannot be supplemented by any of the crossed pool plates.

Solutions of pools 1–11 can be made up by mixing equal volumes of each of the supplements contained in each pool (see Davis *et al.*, 1980 or Maloy *et al.*, 1996). Use 25 ml of the "pool solution" per liter of medium.

Media

LB BROTH

Dissolve the following reagents in 1 liter dH$_2$O then autoclave.

10 g   Tryptone
5 g   Yeast extract
5 g   NaCl

LB AGAR

To make LB plates add 15 g agar per liter before autoclaving.

EBU AGAR

EBU Agar Mix: Combine following ingredients in a large container.

250 g   Bacto Tryptone
125 g   Bacto Yeast Extract
125 g   NaCl
62.5 g   Glucose
375 g   Bacto Agar

25% K$_2$HPO$_4$:

Bring 25 g K$_2$HPO$_4$ to 100 ml with distilled H$_2$O. Autoclave.

1% Uranine:

Bring 1 g Uranine (also known as Fluorescein) to 100 ml with distilled H$_2$O.

Autoclave. Store in a dark bottle or a foil wrapped bottle to avoid exposure to light.

EBU Plates:

Add 37.5 g EBU agar mix to 1000 ml dH$_2$O in a 2 L flask. Autoclave.

After autoclaving, allow the medium to cool to about 50 °C, then add:

10.0 ml of 25% K$_2$HPO$_4$
1.25 ml of 1% Evans Blue
2.5 m of 11% Uranine

Mix thoroughly then pour into sterile petrie dishes.

E MEDIUM

To make 50× E stock

10 g   MgSO$_4$ 7H$_2$0
100 g   Citric acid 1H$_2$0 (Granular)

655 g K_2HPO_4 $3H_20$ (potassium phosphate dibasic)
175 g $NaNH_4HPO_4$ $4H_20$ (sodium ammonium phosphate)
Heat 500 ml deionized water on stirring block but do not boil.
Add the chemicals one at a time. Allow each chemical to dissolve completely before adding the next.
Bring to 1000 ml with deionized water. Allow to cool.
Add about 25 ml chloroform to ensure sterility over long term storage.
Store at room temperature.

To Make E + Glucose Plates

Add 20 ml of 50× E medium to 500 ml deionized water. Autoclave.
Add 15 g Bacto-agar to 500 ml deionized water. Autoclave.
After autoclaving, mix the two containers, then add 10 ml of 20% glucose.
Swirl to mix the solution and pour plates (Table IV).

TABLE IV
ANTIBIOTIC CONCENTRATIONS

Antibiotic	Abbreviation	Final concentration[a]		Stock solution[b]
		Rich media	Minimal media	
Na Ampicillin[c]	Amp or Ap	30 μg/ml	15 μg/ml	9 mg/ml
Chloramphenicol	Cam or Cm	20 μg/ml	5 μg/ml	6 mg/ml
Kanamycin SO_4[d]	Kan or Km	50 μg/ml	125 μg/ml	15 mg/ml
Tetracycline HCl[e]	Tet or Tc	20 μg/ml	10 μg/ml	6 mg/ml

[a] The solid form of the antibiotics or filter sterilized concentrates can be added directly to sterilized media that has been cooled to approximately 55 °C. If kept at 4 °C tetracycline, chloramphenicol, and streptomycin plates are usually good for several months, but kanamycin plates and ampicillin plates may only last for several weeks. Note that when multiple antibiotics are used in together, the concentrations may need to be decreased. A sensitive and resistant control should always be included each time antibiotic plates are used.

[b] For liquid media or for a few plates, a stock solution of the antibiotics can be prepared and stored at −20 °C. Each of these antibiotic stock solutions can be prepared in sterile dH_2O except chloramphenicol which can be dissolved in dimethylformamide. The stock solution is 300× the concentration of antibiotic required in rich medium. An average petrie dish contains about 30 ml medium, so spread 0.1 ml on rich plates.

[c] High level expression of β-lactamase can destroy ampicillin in the medium surrounding the Amp^R colonies, allowing Amp^s satellite colonies to grow. Satellite colonies can be decreased by using 2× ampicillin when selecting for plasmids.

[d] Selection for a Kan^R requires phenotypic expression by incubating in broth for 1–2 generations or plating on nonselective medium, incubating until growth appears, then replica plating on selective medium.

[e] Certain strains are only resistant to half this concentration of tetracycline.

References

Berlyn, M., Low, K., Rudd, K., and Singer, M. (1996). Linkage map of *Escherichia coli* K-12, Edition 9. *In* "*Escherichia coli* and *Salmonella typhimurium*: Cellular and Molecular Biology" (F. Neidhardt, R. Curtiss, J. Ingraham, E. Lin, K. Low, B. Magasanik, W. Reznikoff, M. Riley, M. Shcaechter, and H. Umbarger, eds.), 2nd Edn. ASM Press, Washington, DC.

Davis, R., Botstein, D., and Roth J. (1980). "Advanced Bacterial Genetics," pp. 2–4. Cold Spring Harbor Laboratory, NY.

Maloy, S. (1990). "Experimental Techniuqes in Bacterial Genetics." Jones and Bartlett Publishers, Boston, MA.

Maloy, S. R., Stewart, V. J., and Taylor, R. K. (1996). "Genetic Analysis of Pathogenic Bacteria." Cold Spring Harbor Laboratory, Plainview, NY.

Sanderson, K., Hessel, A., Liu, S. L., and Rudd, K. (1996). The genetic map of *Salmonella typhimruium*, Edition VIII. *In* "*Escherichia coli* and *Salmonella typhimurium*: Cellular and Molecular Biology" (F. Neidhardt, R. Curtiss, J. Ingraham, E. Lin, K. Low, B. Magasanik, W. Reznikoff, M. Riley, M. Shcaechter, and H. Umbarger, eds.), 2nd Edn. ASM Press, Washington, DC.

[6] Localized Mutagenesis

By STANLEY R. MALOY

Abstract

Localized mutagenesis can be used to obtain mutants in genes of interest based on linkage to selectable markers. Mutagens diethylsulfate and hydroxylamine are used to obtain predominantly transition mutations in the DNA either by whole chromosomal mutagenesis or mutagenesis of DNA isolated as purified plasmid or packaged in transducing phage particles. Selectable markers can include those based on auxotrophic requirements, carbon or nitrogen source utilization, or antibiotic resistance markers, such as those encoded in transposons.

Introduction

Antibiotic-resistant transposons have many useful properties as described in Chapter 2. These properties make transposons very useful for the initial genetic analysis of a function. However, most transposon insertions result in complete loss of function, and point mutations that result in leaky, conditional, or altered phenotypes are often useful for detailed characterization of the structure and function of a gene product. Single nucleotide substitutions are particularly valuable for fine-structure genetic analysis.

METHODS IN ENZYMOLOGY, VOL. 421   0076-6879/07 $35.00
DOI: 10.1016/S0076-6879(06)21006-1

The structure and function of gene products are determined by subtle or partially redundant interactions, hence sometimes a single nucleotide substitution is only partially disruptive or "leaky." This may cause a reduction, rather than a complete loss, of enzyme activity. For example, a bacterium with a leaky mutation in a gene that encodes an essential protein (an "essential gene") might form small colonies due to the limited availability of active gene product.

Conditional mutations only produce a mutant phenotype under certain circumstances. Conditional mutants may be affected by environmental conditions or the presence of other mutations. Heat-sensitive (Ts) mutants have a wild-type phenotype at one temperature (such as 30°), and a mutant phenotype at another temperature (such as 42°). Heat-sensitive mutations fall into two classes. First, most Ts mutations affect the folded conformation of a protein: at the permissive temperature the protein folds into a functional conformation, but at the nonpermissive temperature the protein is partially unfolded and typically quickly degraded by cellular proteases. Second, some Ts mutations affect the proper assembly of proteins during synthesis: once synthesized, such temperature-sensitive synthesis (Tss) mutant gene products remain active even at the nonpermissive temperature. In contrast to temperature-sensitive mutations, cold-sensitive mutations result in an active gene product at high temperatures (e.g., 42°), but an inactive product at lower temperatures (e.g., 20° to 30°). Cold-sensitive mutations often affect hydrophobic interactions that mediate protein–protein or protein–membrane associations. Other examples of conditional mutations include suppressor-sensitive mutations, which exhibit a wild-type phenotype in some bacterial strains and a mutant phenotype in others.

Conditional mutations have several useful applications. For example, conditional mutations can be used to identify and characterize the physiological effect of a gene product by shifting to the nonpermissive condition to deplete the gene product, and studying the resulting phenotype as the gene product becomes limiting. Conditional mutations can be used to determine the order of action of gene products in a complex pathway (Jarvik and Botstein, 1973).

Often interesting classes of point mutations cannot be selected for directly, demanding the painstaking screening of thousands of isolated colonies to find the desired mutations following random mutagenesis of the whole genome. Localized mutagenesis provides a much more facile approach to isolate point mutations in a particular locus. Localized mutagenesis can be done in two ways: (1) the entire genome of a donor can be mutagenized and then the desired region of DNA can be selectively transferred to a recipient cell; or (2) a DNA fragment carrying the desired locus can be mutagenized *in vitro*, and then transferred to a recipient cell. A small DNA fragment can be randomly mutagenized *in vitro* by exposing it to a chemical mutagen. In both cases, the mutagenized DNA fragment is brought into an appropriate recipient cell by taking advantage of a linked, selectable genetic marker such as an adjacent chromosomal gene or an antibiotic-resistant transposon insertion. This approach allows localized random mutagenesis of a small, specific region of chromosomal DNA without producing secondary mutations elsewhere on the chromosome. Such localized mutagenesis is especially useful for obtaining rare point mutations in or near a gene of interest (e.g., temperature-sensitive mutations in a gene, mutations in the promoter or operator of a gene, or mutations that affect amino acids at the active site of an enzyme).

Diethylsulfate Mutagenesis

Diethylsulfate (DES) (CH_3CH_2-O-SO_2-O-CH_2CH_3) is an alkylating agent that reacts with guanine to produce ethyl guanine. Mispairing of ethyl guanine can cause G:C to A:T transition mutations. However, in addition to causing mutations directly due to mispairing, the alkylated DNA can cause mutations indirectly by inducing error-prone repair (Drake, 1970).

The donor strain is treated with DES. This procedure can induce mutations anywhere on the chromosome (generalized mutagenesis). Transducing phage is grown on the mutated culture and then used to transfer the desired locus into an unmutagenized recipient. When a population of bacteria is mutated in a liquid culture, each mutant can divide, yielding many siblings with the same mutation. Genetic and biochemical characterization of mutants can be a lot of work, so it is important to be confident that you are not wasting time repeating work on siblings with the identical mutation. In order to avoid the characterization of siblings, usually only one mutant with a specific phenotype is saved from each culture.

Hydroxylamine Mutagenesis

In addition to *in vivo* mutagenesis as described for DES, hydroxylamine (NH_2OH) can be used to mutagenize DNA *in vitro*. When used *in vitro*, hydroxylamine reacts with cytosine, converting it to a modified base that pairs with adenine. This has two consequences: (1) hydroxylamine only produces G:C to A:T transitions, and (2) mutations induced by hydroxylamine cannot be reverted with hydroxylamine. (If hydroxylamine is used *in vivo*, the resulting DNA damage induces error-prone repair, which can result in a wide variety of mutations.) In addition, hydroxylamine gives a high ratio of mutagenic to lethal events.

In contrast to many other mutagens, hydroxylamine can mutagenize DNA packaged inside of phage heads. This allows localized mutagenesis of transducing particles (Hong and Ames, 1971). A phage lysate containing transducing particles is mutagenized with hydroxylamine *in vitro*, and then cells are infected selecting for a marker linked to the desired locus. When the transducing fragment is recombined onto the chromosome, only the small, localized region carried on that transducing fragment is mutagenized. When isolating mutants from independently mutagenized transducing fragments, problems with isolation of siblings are avoided. The extent of mutagenesis of the phage particles can be monitored indirectly by following decrease in phage titer ("killing") due to mutations in essential phage genes, or directly by following the increase in clear-plaque phage mutants in the lysate (Fig. 1).

Following transduction and selection for the linked marker, the resulting transductants are screened for point mutations with the desired phenotype (that is, the "unselected" marker). The frequency of such mutations will depend on how heavily the transducing lysate was mutagenized and how closely linked the selected and unselected markers are.

Essentially any marker with a selectable phenotype can be used for localized mutagenesis. For example, the selected marker can repair an auxotrophic mutation (Fig. 2). However, many genes do not have closely linked, easily selectable genetic markers. In these cases, a linked transposon can be used as the selectable marker (Fig. 3). For example, a transducing

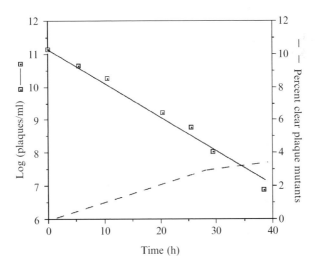

FIG. 1. The effect of exposure time to hydroxylamine on the survival of P22 phage particles and the appearance of clear-plaque mutant phages.

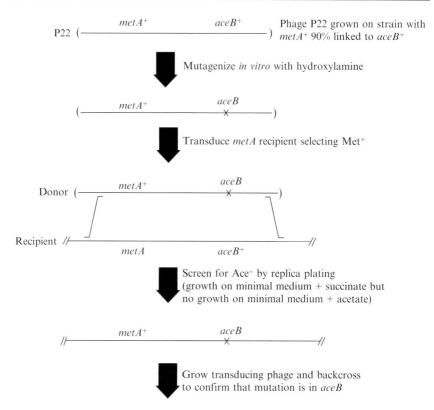

Fig. 2. Use of auxotrophic marker ($metA(+)$) for localized mutagenesis of a linked gene ($aceB$) by P22 co-transduction.

lysate grown on a strain with a Tn*10* insertion closely linked to some particular gene can be used to bring the linked gene into a recipient cell by selection for the antibiotic resistance (Tet^R) encoded by Tn*10*. Using this approach, it is possible to realize localized mutagenesis of essentially any region of the chromosome.

By using this technique it is even possible to heavily mutagenize a small region of the chromosome without mutagenizing the rest of the chromosome. Localized mutagenesis is especially useful for obtaining rare cis-dominant regulatory mutants linked to a gene (Hahn and Maloy, 1986) or rare types of mutations in a structural gene (Dila and Maloy, 1987).

Hydroxylamine can also be used to mutate purified plasmid DNA *in vitro*. The procedure for mutagenesis of plasmids is essentially identical to the procedure for phage, except after mutagenesis the hydroxylamine is

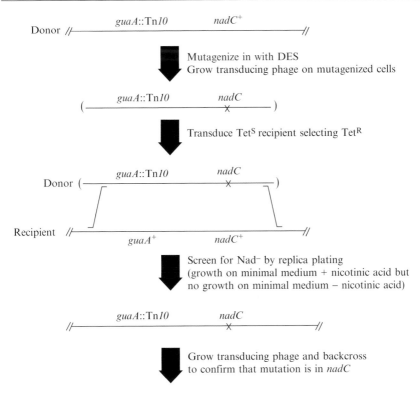

FIG. 3. Use of linked transposon marker (guaA::Tn10) for localized mutagenesis of a linked gene (nadC) by P22 co-transduction.

removed by dialysis prior to transforming cells (Klig *et al.*, 1988). The transformants are selected on antibiotic plates, and then the colonies are replica plated onto appropriate media to screen for the desired mutations. It is a good idea to assay mutagenesis of another plasmid gene as a control. Some plasmids are more difficult to mutagenize than others. This may be due to secondary structure in the DNA, which strongly inhibits mutagenesis by hydroxylamine (Drake, 1970). If this is a problem, it may be necessary to partially denature the DNA by doing the mutagenesis at a higher temperature (Humphreys *et al.*, 1976).

Mutagenesis with Diethylsulfate

Caution: Mutagens are potential carcinogens. Wear gloves and do *not* mouth pipet. Dispose of all waste containing the mutagen in appropriate biohazard waste containers.

1. Add 50 μl of DES to a 10-ml screw-capped test tube containing 2.5 ml of minimal medium with no carbon source. Tighten the cap. Vortex the tube, and then place it in a 37° water bath for 10 min to form a saturating solution of DES.
2. Add 50 μl of an overnight culture (about 10^9 cells per milliliter) to the aqueous phase. Do not shake the tube when adding the cells. Also add 50 μl of the overnight culture to a control tube containing 2.5 ml of minimal medium without DES.
3. Incubate 50 min at 37° without shaking.
4. Remove 50 μl and subculture into 2 ml of rich medium.
5. Grow several generations at 37°. This allows the treated cells to recover from the mutagenesis and mutant chromosomes to segregrate from nonmutant sister chromosomes.
6. Dilute the mutagenized and control cultures in sterile 0.85% NaCl. The cultures should be diluted sufficiently to yield 1000 to 2000 bacteria per milliliter.
7. Plate 0.1 ml of each dilution on a rich medium plate. Incubate at 37° overnight.
8. Replica plate onto appropriate plates to screen for the desired phenotype.
9. Incubate the plates overnight at 37°. (The NCE + lactose plates may take 2 to 3 days.)
10. Examine the plates. It is important to pay close attention to the ratio of mutants obtained. Note the number of colonies on each plate and the number of colonies with the desired phenotype. An effect control for the efficiency of mutagenesis is to screen for auxotrophic mutants (colonies that grow on the rich medium plate but not a minimal medium plate).

Hydroxylamine Mutagenesis *In Vitro*

Caution: Mutagens are potential carcinogens. Wear gloves and do *not* mouth pipet. Dispose of all waste containing the mutagen in appropriate biohazard waste containers.

1. Grow a generalized transducing phage on a sensitive recipient strain with a marker linked to the region of interest. It is important to use a high-titer lysate. If necessary, the phage can be concentrated as described in Steps 6 and 7.
2. Add the following solutions to two separate sterile test tubes:
   Mutagenesis tube: phosphate-EDTA buffer, 0.40 ml; sterile dH$_2$O, 0.60 ml; hydroxylamine, 0.80 m; 1 *M* MgSO$_4$, 0.02 ml
   Control tube: phosphate-EDTA buffer, 0.40 ml; sterile dH$_2$O, 1.40 ml; 1 *M* MgSO$_4$, 0.02 ml

3. Add 0.2 ml of the generalized transducing phage lysate to each tube.
4. Incubate 24 to 48 hrs in a 37° incubator. Remove samples at 0 time and every 4 to 8 hr. Dilute 10 μl into 1 ml of cold Luria broth-salt-EBTA (LBSE). Titer each sample on a sensitive recipient strain to determine the decrease in viable phage and the proportion of plaque mutants. (Remember to take the decrease in phage titer into account when planning dilutions. The plaque mutants will account for only a few percentage points of the total phage.)
5. Plot the plaque-forming units per milliliter versus time on semi-log paper. Predict when killing will reach 0.1 to 1.0% survivors (usually 24 to 36 hr).
6. At the last time point, remove an aliquot for titering, and then centrifuge the rest at 15,000 rpm for 2 hr at 4° in the SS34 rotor.
7. Pour off the supernatant. Overlay the phage pellet with 1 ml of cold LBSE. Place at 4° overnight, followed by occasional, gentle swirling to resuspend the pellet. (Do not vortex or attack the pellet with a pipet.)
8. Dilute the mutagenized phage 1/10 in T2 buffer for transductions. Transduce a sensitive recipient with the mutated phage as shown below.
      Plate A: 0.1 ml bacteria; cell control
      Plate B: 0.2 ml phage; phage control
      Plate C: 0.1 ml bacteria; 0.05 ml phage
      Plate D: 0.1 ml bacteria; 0.1 ml phage
      Plate E: 0.1 ml bacteria; 0.2 ml phage
9. Spread each sample on a plate that selects for the linked genetic marker. Incubate the plates upside-down at 37°. (Note that selection for certain antibiotic resistance phenotypes, particularly Kanamycin resistance, demands phenotypic expression prior to plating on the selective medium.)
10. Check the plates. There should be no growth on either of the control plates. Replica plate the transduction plates onto media to screen for the desired mutants in the unlinked locus.

Reagents

LBSE

1. Mix 100 ml Luria-Bertani broth (LB), 0.2 ml 0.5 M EDTA, and 5.85 g NaCl.
2. Autoclave, and then store at 4°.

Phosphate-EDTA Buffer (0.5 M KPO₄ pH 6, 5 mM EDTA)

When exposed to oxygen, hydroxylamine solutions form by-products (peroxides and free radicals) that are toxic to bacteria. This nonspecific toxicity is decreased by EDTA.

1. Dissolve 6.81 g of KH_2PO_4 in 70 ml of dH_2O on a stirrer.
2. Bring to pH 6 with 1 M KOH.
3. Bring to 99 ml with dH_2O.
4. Add 1 ml of 0.5 M EDTA.
5. Autoclave.

Hydroxylamine/NaOH (Prepare Fresh)

1. Mix 0.175 g of hydroxylamine (NH_2OH) and 0.28 ml 4 M NaOH.
2. Bring to 2.5 ml with sterile dH_2O.

References

Dila, D., and Maloy, S. (1987). Proline transport in *Salmonella typhimurium: putP* permease mutants with altered substrate specificity. *J. Bacteriol.* **168,** 590–594.

Drake, J. (1970). "The Molecular Basis of Mutation," pp. 152–155. San Francisco: Holden-Day.

Hahn, D., and Maloy, S. (1986). Regulation of the put operon in *Salmonella typhimurium*: Characterization of promoter and operator mutations. *Genetics* **114,** 687–703.

Hong, J., and Ames, B. (1971). Localized mutagenesis of any specific small region of the bacterial chromosome. *Proc. Natl. Acad. Sci. USA* **68,** 3158–3162.

Humphreys, G., Willshaw, G., Smith, H., and Anderson, E. (1976). Mutagenesis of plasmid DNA with hydroxylamine: Isolation of mutants of multi-copy plasmids. *Mol. Gen. Genet.* **145,** 101–108.

Jarvik, J., and Botstein, D. (1973). A genetic method for determining the order of events in a biological pathway. *Proc. Natl. Acad. Sci. USA* **70,** 2046–2050.

Klig, L., Oxender, D., and Yanofsky, C. (1988). Second-site revertants of *Escherichia coli trp* repressor mutants. *Genetics* **120,** 651–655.

Maloy, S., Stewart, V., and Taylor, R. (1996). "Genetic analysis of pathogenic bacteria." Plainview, NY: Cold Spring Harbor Laboratory Press.

Roth, J. (1970). Genetic techniques in studies of bacterial metabolism. *Methods Enzymol.* **17,** 1–35.

[7]   Generation of Deletions and Duplications Using Transposons as Portable Regions of Homology with Emphasis on Mud and Tn10 Transposons

By Kelly T. Hughes

Abstract

In bacteria complementation and dominance testing requires the establishment of a diploid state for the gene of interest. In addition, it is often desirable to characterize reporter fusion constructs in strains with both the reporter fusion and an intact gene copy present in single copy. Transposons provide portable regions of homology to facilitate construction of targeted chromosomal rearrangements such as deletions and duplications. The properties of the large Mud transposons, MudA and MudB allow for the direct duplication and deletion of virtually any region of the *Salmonella enterica* chromosome between the points of two Mud insertions in a simple bacteriophage P22 transductional cross. Furthermore, duplication construction will be described for the generation of strains with a *lac* operon transcriptional fusion or *lacZ* gene translational fusion to any gene of interest at the join-point of the duplication with a second intact copy of the gene of interest located in tandem single copy in the same chromosome. In addition, methods for generation of tandem chromosomal duplications using transposon Tn10 as portable regions of homology are presented. These allow construction of strains duplicated for any gene of interest in tandem, single copy on the chromosome to allow for complementation and dominance testing for alleles for virtually any gene.

Introduction

Transposon insertions or targeted DNA cassettes with selectable markers can be isolated in every location in a given chromosome. Multiple DNA elements of identical or near-identical sequence can act as portable regions of homology for recombination to occur between them. Deletions and duplications can be formed by unequal crossing-over between DNA sequences that are in direct repeats. (Duplications of adjacent sequences are called tandem duplications.) When duplications occur via intramolecular recombination (i.e., between sequences on the same DNA molecule), they probably result from recombination between sister chromosomes soon after the DNA has replicated. The following figures show the two

METHODS IN ENZYMOLOGY, VOL. 421   0076-6879/07 $35.00

consequences of unequal crossovers, drawn as half-crossover events to emphasize how duplications arise versus how deletions arise.

Because transposons provide portable regions of homology that can be inserted at specific regions of the chromosome, two copies of a transposon inserted in the same orientation on the chromosome can provide the direct repeats needed to duplicate or delete any desired region of the chromosome. Some deletions are not viable if the deleted material includes a gene essential for growth unless the gene is provided artificially on a plasmid or another location in the chromosome. Examples are shown below. The orientation of the transposon insertion is indicated by the direction of the arrowheads. These duplications will be unstable because recombination between the direct repeats can delete the duplicated region (a process called segregation).

Generation of tandem chromosomal duplications can be used to set up complementation and dominance tests without the problems associated with copy number effects using plasmid-based duplication systems. The recovery of individual alleles tested following segregation is excellent proof that both markers were present during the complementation or dominance assay.

The use of Mud and Tn*10* elements for the generation of duplications and deletions in genetic analysis is described in this chapter.

General Considerations

Introduction

*The Transposable Elements, Tn*10 *MudA and MudB*

Transposons Tn*10* is 10 kbp in length while the MudA and MudB elements are 37 kbp in size. Thus each transposon provides a large region for homologous recombination to occur between multiple elements. Tn*10* and MudA/B encode resistances to tetracycline (Tc(R)) and ampicillin (Ap(R)), respectively, to be used as selective markers. The MudA transposon is a derivative of transposon MudI designed for the construction of *lac* transcriptional (*lac* operon) fusions to genes of interest (Casadaban and Cohen, 1979; Groisman, 1991; Hughes and Roth, 1984). The MudB transposon is a derivative of transposon MudII designed for the construction of *lacZ* translational (*lacZ* gene) fusions to genes of interest (Casadaban and Chou, 1984; Groisman, 1991; Hughes and Roth, 1984). A problem associated with the MudI and MudII transposons was that they encoded Mu transposase. Thus selection for increased *lac* expression using a strain with a MudI or MudII fusion resulted in transposition of the Mud elements into a new chromosomal location resulting in higher Lac expression. The MudA and MudB elements are identical to MudI and MudII, respectively except the transposase genes in MudA and MudB contains amber mutations. For MudA/B, transposition occurs in strains carrying an amber suppressor mutation, but not in strains lacking an amber suppressor mutation. This allowed for controlled transposition of the MudA/B elements. In strains with an amber suppressor, MudA/B are inherited primarily by transposition while in a strain lacking an amber suppressor mutation MudA/B are inherited by homologous recombination.

Transposon Tn*10* is typically used to generate deletions and duplications on the order of 1–30 kbp in length (Chumley and Roth, 1980; Chumley *et al.*, 1978; Hughes *et al.*, 1983). The MudA/B transposons are typically used to generate duplications on the order of one to hundreds of kbp in length. A related Mud element, MudP22, has been used to generate a set of 11 strains of *Salmonella enterica* that in total duplicate the entire chromosome (Camacho and Casadesus, 2001).

Use of Phage P22 to Transduce Tn10 and MudA/B

Transposon Tn*10* transposes at a low frequency 10^{-5} (Kleckner, 1983). In *Salmonella enterica*, transposon Tn*10* is introduced on a P22 phage that is unable to grow in the recipient cell (Kleckner *et al.*, 1991). Infection of 1 ml of an overnight culture of *S. enterica* cells (10^9 cells) with a 100-fold excess of defective P22 particles that carrying Tn*10* will yield 10^6 independent Tc(R) transposition events. These can then be moved from strain to strain by generalized transduction and are inherited by homologous recombination with only a low probability of transposition 10^{-5}.

Phage P22 *HT105/1* is a Salmonella phage that performs high-frequency generalized transduction. That is, when P22 *HT105/1* multiplies in a cell, 50% of the packaged fragments are of the host chromosome within its particles. P22 is 42 kbp in length (Pedulla *et al.*, 2003). However, due to exonuclease degradation upon injection of a transduced fragment, markers that are more than 32 kbp apart are co-transduced at a very low efficiency (<1%). The MudA/B elements are close in size to the limit of what P22 can package (37 and 36 kbp, respectively [Groisman, 1991]). P22-mediated transduction of MudA/B elements and inheritance by homologous recombination requires two transduced fragments, each carrying a part of the Mud element from the donor cell. Inheritance by homologous recombination requires three recombinational exchanges, one between Mud sequences of the individual fragments and two between the ends of the composite Mud element and the recipient chromosome.

Generation of Deletions and Duplications Using Tn10 or Mud by P22 Transduction Tn10

Insertions of Tn*10* that flank a gene of interest can be used as portable regions of homology to delete or duplicate that gene. First, Tn*10* insertions are isolated that are linked to the gene of interest by P22 cotransduction. The cotransduction frequency between Tn*10* and the gene of interest can be used to estimate between the insertion and the gene of interest (Sanderson and Roth, 1988). If desired, the sequence of the chromosome DNA flanking the Tn*10* insertion can be obtained to determine the exact position in the chromosome. Duplications and deletions between Tn*10* elements require that the elements be inserted in the same orientation in the chromosome. The simplest method to determine the orientation of Tn*10* in the chromosome is by PCR using an oligo that hybridizes to unique sequences within Tn*10* and oligos to genes in the region where the Tn*10* is inserted. Tn*10* is a composite transposon that includes 7 kbp of DNA encoding tetracycline resistance flanked by nearly identical 1.4 kbp IS*10* elements in inverted orientation with respect to each other (Kleckner, 1983). Thus the primer for Tn*10* is designed using the unique DNA sequence that encodes tetracycline resistance and not using IS*10* sequence that is duplicated.

Once Tn*10* insertions have been identified that flank the gene of interest and are in the same orientation on the chromosome, they can be used to generate deletions and duplications between the points of insertion as diagrammed in Fig. 1. The selection for duplications requires at least one

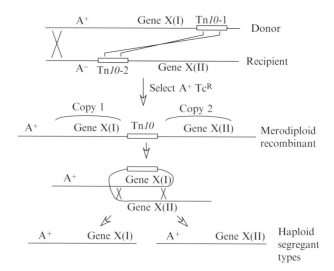

Fig. 1. Use of Tn*10* insertions flanking a gene of interest to generate a chromosomal duplication in which each copy of the duplicated region contains an allele of the gene of interest. The duplication carries a hybrid Tn*10* at the duplication join-point and is maintained by selection for Tn*10*-encoded Tc(R). When selection for the duplication is removed by growth in the absence of added Tc, recombination between the duplicated regions occurs, yielding haploid segregant types containing only one of the alleles of the gene of interest.

selectable marker (A(+) in Fig. 1) to be located near the genes of interest. Then, selection for that marker (A(+)) and Tc(R) using a donor strain carrying one Tn*10* flanking the gene of interest (gene X) on the side opposite of gene X as the selected marker (Tn*10*-1) can be used to generate a duplication if the recipient has a second Tn*10* inserted (Tn*10*-2) on the side of gene X proximal to recessive marker (A(−)). If the Tn*10* insertions are reversed in the donor and recipient, the same selection (A(+) Tc(R)) yields deletions of gene X as diagrammed in Fig. 2.

Mud

The large size of MudA/B elements relative to the size of DNA packaged by P22 allows direct selection for deletions and duplications of the region of the chromsome between two Mud insertions that are inserted in the same orientation on the chromosome. P22 transducing lysates are prepared on the two strains that each carry a Mud insertion in a different location of the chromosome. The two P22 lysates are then checked to determine which dilution of each lysate will yield equal numbers of Ap(R) transductants. Equal volumes of dilutions that give equal numbers of Ap(R) transductants are then mixed to produce the mixed lysate for the experiment. When the mixed lysate is used to transduce a recipient to Ap(R), the composite Mud element in the recipient can be generated from either the two halves of an original insertion, Mud-1 or Mud-2, or two halves (5′-Mud-1 recombined with 3′-Mud-2 or 5′-Mud-2 recombined with 3′-Mud-1). The hybrid Mud elements are flanked by chromosomal DNA from the different regions of the chromosome corresponding to the insertion sites of the original donor strains Mud-1 or Mud-2. When these hybrid elements are inherited by homologous recombination, three things can occur. If the donor Mud insertions are inserted in the same orientation in the chromosome, inheritance of

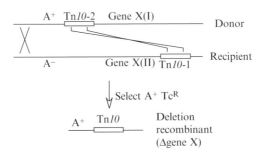

FIG. 2. Use of Tn*10* insertions flanking a gene of interest to generate a chromosomal deletion in which the region of the chromosome between the sites of insertion of the two Tn*10* insertions, including the gene of interest, is deleted.

the hybrid Mud element leads to duplication or deletion of the region of the chromosome between the original points of insertion (see Fig. 3). If the deleted region contains essential genes, deletion recombinants are not viable. If the donor Mud insertions are inserted in the opposite orientation in the

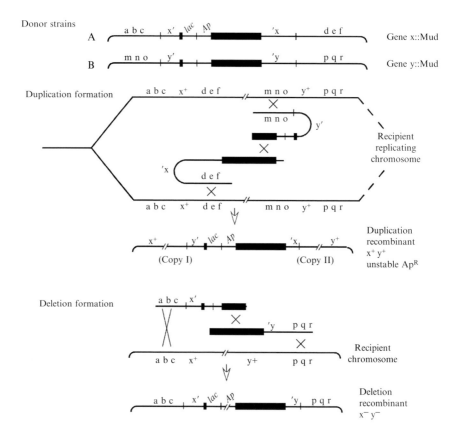

FIG. 3. Recombination between different Mud fragments results in the formation of a duplication or deletion recombinant between the original chromosomal Mud insertion sites. Donor strains A and B carry Mud insertion insertions in hypothetical gene x and gene y, respectively. Both Mud insertions are in the same orientation in the chromosome. When P22 lysates grown on strain A and strain B are mixed and the resulting mixed lysate is used to transduce a recipient to Mud-encoded Ap(R), four different recombinant types can arise. Two recombinant types are the parental insertions in gene x or gene y. Two additional recombinant types result from transduction of two Mud fragments, one from each of the different parent insertion mutants. Depending on which combination of parent fragments is transduced into a recipient, recombination of the hybrid Mud fragment into the recipient chromosome will lead to duplication or deletion events of the chromosomal material between the points of insertion; however, deletions of some regions may not be recovered due to deletion of essential genes or chromosomal DNA (such as the replication terminus).

chromosome, inheritance of the hybrid Mud element leads to a broken chromosome and presumably cell death.

Orientation of MudA is easily determined using a lactose utilization indicator such as X-gal. MudA makes *lac* operon fusions. If MudA inserts in a gene of interest in the orientation where the *lac* operon is transcribed from the gene's promoter the cells will be Lac$^+$. If MudA inserts in a gene of interest in the orientation where the *lac* operon is not transcribed from the gene's promoter the cells will be Lac$^-$. MudB makes *lacZ* translation fusions. Insertions have to be in the correct orientation and reading frame to be Lac$^+$. Thus, with MudB, the orientation of the Lac$^-$ insertions cannot be determined by phenotype and requires PCR amplification between sequence at one end of MudB and nearby chromosomal sequence.

The Plan of the Experiment

Mud

The isolation of *lac* operon and *lacZ* gene fusions in *S. enterica* is routinely done using transposons MudJ (MudI1734) and MudK (MudII1734), respectively (Groisman, 1991; Hughes and Roth, 1988). These transposons are much smaller than MudA (37 kbp) and MudB (36 kbp). They utilize a Km cassette as a selectable marker and lack the Mu transposase genes. Because of the smaller size of MudJ and MudK (11 and 10 kbp, respectively), these elements can by transduced within a single P22 phage particle and inherited by homologous recombination. Transposition of MudJ or MudK elements is usually performed using a donor strain with the MudJ or MudK insertion located in the chromosome near the Mu transposase genes. When a P22 transducing lysate is prepared on the MudJ/K donor strains, both the MudJ/K and adjacent Mu transposase genes can be packaged within the same transducing phage particle. When this particle infects a recipient cell, the Mu transposase genes are expressed and transposase acts on the ends of the adjacent MudJ/K element to allow transposition into the recipient chromosome. The DNA containing the transposase genes is degraded and no further transposition occurs. The method is referred to as transitory cis-complementation (Hughes and Roth, 1988). To generate duplications with MudJ or MudK insertions it is easiest if they are first converted to MudA or MudB insertions, respectively. Fortunately, the DNA sequence that flanks the MudA and MudJ fragments and the sequence that flanks the MudB and MudK fragments are identical and therefore interconvertible. A P22 lysate prepared on a MudA insertion in F plasmid DNA is used to transduce a recipient strain that carries a MudJ insertion in any gene of interest selecting for MudA-encoded Ap(R) and screening for loss of MudJ-encoded Km(R). Similarly, a P22 lysate prepared on a MudB insertion in F plasmid DNA is used to transduce a recipient strain that carries a MudK insertion in any gene of interest selecting for MudB-encoded Ap(R) and screening for loss of MudK-encoded Km(R). This will convert either a MudJ to a MudA or a MudK to a MudB that can then be used to generate Mud-directed duplications and deletions.

Once MudA or MudB insertions are isolated in the gene of interest, P22 lysates are prepared on Mud insertions to be used for generating a deletion or duplication between the insertion points of the different Mud insertions in the donor strains. This is particularly useful for making strains that carry a *lac* transcriptional or *lacZ* translational fusion to a gene of interest and a second intact copy of the gene of interest. If the gene is autoregulatory, then it is critical that assays to characterize the transcriptional or translational regulation of a given gene carry an intact copy of that gene in single copy in the chromosome to avoid artifacts that can be associated with plasmid copy number effects when trying to complement with a plasmid-expressed gene.

P22 lysates are prepared on the two donor Mud insertion mutants. These individual lysates are then used to transduce a recipient strain to mud-encoded Ap(R) to estimate the phage titer. Typically a standard lysate is diluted 100-fold and used to transduce wild-type strain LT2 to Ap(R). Once the number of Ap(R) transductants of each lysate is determined, diluted phage lysates are mixed such that the mixture contains equal numbers of Mud-transducing particles. Since Mud is inherited by two-fragment transduction, a two-fold increase in the phage concentration results in a 4-fold increase in the number of Ap(R) transductants. If 0.1 ml of a 10^{-2} dilution of lysate #1 gives 30 Ap(R) transductants and a 0.1 ml of a 10^{-2} dilution of lysate #2 gives 120 Ap(R) transductants, then equal volumes of a 10^{-2} dilution of lysate #1 and a 2×10^{-2} dilution of lysate #2 are mixed to give equal numbers of Mud transducing particles from the different donor strains in the mixed lysate. The mixed lysate is then used to transduce a recipient cell to Ap(R). If the deletion is viable and the duplication is small (<5% the size of chromosomal size), one half of the Ap(R) transductants are of the parental type from lysate #1 or lysate #2, one fourth are duplicated and carry wild-type copies of the genes in which the Mud insertions were originally located, and one fourth are deleted for the region between the original sites of insertion and are expected to have at least a double mutant phenotype associated with each original insertion mutant plus any phenotype associated with the loss of the intervening chromosomal DNA. If the deletion is not viable and the duplication is small, two thirds of the Ap(R) transductants are of the parental type from lysate #1 or lysate #2, and one third will have the duplicated phenotype (Hughes and Roth, 1985). For large duplications the frequency of duplication recombinants relative to the parental recombinant types will vary with size and location on the chromosome. Also, duplications that include the terminus of replication for the chromosome are not viable.

Tn10

First a P22 transducing lysate is prepared on a large pool of independent Tn10 insertion mutants. To ensure good saturation mutagenesis, the number of insertion mutants in the pool should be about 10 times the number of kbp in the genome. For *S. enterica*, which is about 5000 kbp, a pool of at least 50,000 independent Tn10 insertion mutants is used. To obtain a Tn10 insertion near a gene of interest, a mutant in the gene of interest with a discernible phenotype is required. The P22 transducing lysate prepared on the Tn10 pool is used to transduce the mutant to Tn10-encoded Tc(R) and followed by a screen for repair of the mutant allele by cotransduction. P22 will package 42kbp of DNA. The co-transduction frequencies between a given Tn10 insertion and the gene of interest will range from >90% for insertions within a few kbp from the gene of interest to a few percent for insertions 25–30 kbp away. About 500 Tc(R) transductants will need to be screened to obtain a Tn10 insertion linked to the gene of interest. Once linked insertions are isolated, the cotransduction frequency is determined. A P22 lysate is prepared on each linked Tn10 insertion mutant and used to transduce a strain mutant in the gene of interest to Tc(R). The percentage of Tc(R) transductants that repair the mutant allele to the wild-type phenotype is the cotransduction frequency. This frequency can be used to estimate the physical distance between the site of Tn10 insertion and the gene of interest by a modification of the Wu formula (Sanderson and Roth, 1988). Once linked Tn10 insertions are obtained and cotransduction frequencies are obtained, individual insertions are screened for location clockwise or counterclockwise to the gene of interest and orientation in the chromosome by PCR using primers unique to non-IS10 DNA within Tn10 and DNA in the vicinity of where the transposon is located based on the cotransduction frequency.

Once Tn10 insertions are characterized for orientation in the chromosome and position, clockwise or counterclockwise, on the chromosome relative to the gene of interest, deletions and duplications between the Tn10 elements can be constructed. A selectable marker near the gene of interest is required as diagrammed in Fig. 1. If none is available, another transposon insertion or targeted cassette insertion with a different selectable marker such as Km or Cm can be used. Duplication recombinant will lose Tn10 and its associated Tc(R) if grown in medium lacking Tc. The frequency that this occurs depends on the size of the duplicated region and therefore the amount of duplicated DNA available for homologous recombination. Thus, growth of duplicated strains in the presence of Tc will maintain the duplication in the culture.

Procedure 1: Growing P22 lysates.

Solutions

LB (Luria broth): Per liter deionized water, 10 g tryptone, 5 g yeast extract, 5 g NaCl.

Ex50 salts: 50% D-glucose.

Sterile saline: Per liter deionized water, 8.5 g NaCl.

Top agar: Per liter deionized water, 10 g tryptone, 7 g agar.

P22 broth: 200 ml LB, 2 ml Ex50 salts, 0.8 ml 50% D-glucose, 10^7–10^8 plaque forming units (pfu)/ml.

P22 transducing phage: P22 HT/*int*.

P22 HT/*INT* LYSATE PREPARATION. Grow a P22-sensitive host strain to saturation in LB. Make serial dilutions of a P22 lysate in sterile saline and plate 0.1 ml of diluted phage with 0.1 ml of cell culture in 3 ml top agar on a LB agar plate (12 g agar per liter). Pick a single plaque with a pasteur pipette and inoculate a 1 ml LB saturated culture of a sensitive strain. Add 4 ml of P22 broth that does not have added P22 and grow with shaking at 37° for 5 or more hours (lysates left over the weekend will work, but usually an all-day or overnight incubation period is used for convenience). Titer the resulting lysate and use it to prepare a working stock of P22 broth. For all future lysates add 4 ml of P22 broth to 1 ml LB saturated culture of a sensitive strain and grow 5–36 hours at 37° with shaking. Pellet cells by centrifugation (10 min full speed in a table top centrifuge or for larger volumes spin 5 min at 8000 rpm in a SS34 rotor). Decant the supernatant into a sterile tube, add ChCl₃ and vortex to sterilize. Store at 4°.

Procedure 2: Transposition of MudJ/K into the chromosome of *Salmonella enterica*. Isolation of MudJ/K fusions to genes of interest by replica printing.

Materials

P22 HT/*int* transducing lysates on *S. enterica* MudJ and MudK donor strains (TH2142 = *hisD9953*::MudJ *his-9944*::MudI and TH2145 = *hsiD1284*::MudK *his-9944*::MudI).

P22 HT/*int* transducing lysates on *S. enterica* strains carrying MudA or MudB insertions in F plasmid DNA for cassette replacement experiments (TH331 = F′128 *zzf-1066*::MudA/ *proAB47 pyrB64* and TH1123 = F′152 *zzf-1093*::MudB/ *nadA56*).

Recipient wild-type *S. enterica* strain.

1. Start a 1 ml overnight culture of the wild-type *Salmonella enterica* strain. Grow overnight with aeration at 37°.

2. First do a test cross to determine the number of MudJ or MudK insertions obtained per 0.1 ml of diluted phage stock. Dilute the MudJ and MudK donor lysates 10^{-2}, 10^{-3}, and 10^{-4}. In a sterile tube, mix 0.1 ml of cells from the overnight culture with 0.1 ml of phage grown on the MudJ or MudK donor. Some of the phage particles will inject Mud DNA and the adjacent Mu transposase genes from the linked MudI insertion into your recipient cells. For experiments that will require replica printing to screen for insertions in the desired genes, 300–500 colonies per plate is a good working number. The size of the target will determine the frequency at which a MudJ or MudK insertion is obtained. For a 1 kbp gene target, at least 5000 Km(R) will need to be screened to obtain each MudJ insertion and 10,000 for each MudJ inserted in the correct orientation to place the *lac* operon under control of the promoter of the gene into which MudJ has inserted.

3a. Start a fresh 20 ml overnight culture of the *Salmonella enterica* recipient strain. Grow overnight with aeration at 37°. For MudJ insertions that require a screening step by replica printing add 11 ml of cells to 11 ml of diluted phage using the dilution that gave 300–500 Km(R) transductants per plate. Let sit 1 hour at room temperature to allow for phenotypic expression of Km(R). Plate 0.2 ml onto each of 100 LB plates with added Km (50 µg/ml). Incubate overnight at 37°.

3b. Since MudK insertions have to be in the correct orientation and reading frame, more colonies will have to be screened to obtain LacZ protein fusions (translational fusions) to genes of interest. Start a fresh 60 ml overnight culture of the *Salmonella enterica* recipient strain. For MudK insertions that require a screening step by replica printing, add 55 ml of cells to 55 ml of diluted phage using the dilution that gave 300–500 Km(R) transductants per plate. Let sit 1 hour at room temperature to allow for phenotypic expression of Km(R). Plate 0.2 ml onto each of 500 LB plates with added Km (50 µg/ml). Incubate overnight at 37°.

4. Replica-print LB-Km transduction plates to the medium that will identify insertions in the gene(s) of interest. For example, to obtain MudJ/K insertions in the histidine biosynthetic operon each plate will be replica printed to two minimal salts glucose medium plates. The first plate will have added Km (50 µg/ml) and the second plate will have added Km (50 µg/ml) and added histidine (0.1 m*M*). Any colonies that require histidine for growth in minimal medium are picked as potential insertions in the histidine biosynthetic operon. The colonies are then screened on lactose indicator media for functional *lac* transcriptional fusions for MudJ or *lacZ* translational fusions for MudK. Standard lactose indicator media are X-gal (40 µg/ml), MacConkey-lactose or triphenyltetrazolium chloride-lactose (TTC-Lac) media (Maloy *et al.*, 1996). The different lactose indicator media require different levels of β-galactosidase (β-gal) activity to

produce a Lac$^+$ phenotype. Media with X-gal requires the lowest levels of β-gal activity to produce a Lac$^+$ phenotype and TTC-Lac the highest.

5. Read replica prints and using a toothpick, pick the MudJ/K insertions in the gene of interest and isolate P22-sensitive mutants using green indicator plates (Maloy et al., 1996).

6. Prepare P22 transducing lysates on MudJ/K insertion mutants and use these to transduce the parent strain to Km(R). Again screen for the mutant phenotype and isolate P22-sensitive mutants using green indicator plates. About 5% of MudJ/K insertion mutants have multiple Mud insertions in the cell that need to be separated by transducing out into a clean background.

Procedure 3: Transposition of MudJ/K into the chromosome of *Salmonella enterica*. Isolation of MudJ/K fusions to genes by linkage to a known marker.

Materials

P22 HT/*int* transducing lysates on *S. enterica* MudJ and MudK donor strains (TH2142 = *hisD9953*::MudJ *his-9944*::MudI and TH2145 = *hsiD1284*::MudK *his-9944*::MudI).

P22 broth: LB with added minimal medium salts such as E-salts (Maloy et al., 1996), 0.2% glucose, and P22 HT/*int* transducing phage at 5×107 pfus (plaque forming units) per ml.

Recipient *S. enterica* strain with marker near gene of interest. This can be an auxotrophic gene near the gene to be targeted or a linked drug resistance marker such as a Tn10dTc, Tn10dCm, or targeted drug-resistant cassette inserted near the gene of interest. As an example, a linked Tn10dTc insertion will be used.

1. Start a 1 ml overnight culture of the *Salmonella enterica* strain with a Tn10dTc linked to the gene of interest by P22 transduction by at least a 50% cotransduction frequency. Grow overnight with aeration at 37°.

2. First do a test cross to determine the number of MudJ or MudK insertions obtained per 0.1 ml of diluted phage stock. Dilute the MudJ and MudK donor lysates 10^{-1}, 10^{-2} and 10^{-3}. In a sterile tube, mix 0.1 ml of cells from the overnight culture with 0.1 ml of phage grown on the MudJ or MudK donor. Some of the phage particles will inject Mud DNA and the adjacent Mu transposase genes from the linked MudI insertion into your recipient cells. For experiments that use a linked marker, a dilution that yields 5000–10,000 colonies per plate is a good working number.

3a. Start a fresh 3 ml overnight culture of the *Salmonella enterica* recipient strain. Grow overnight with aeration at 37°. For MudJ insertions using a linked Tn10dTc marker, add 2.2 ml of cells to 2.2 ml of diluted phage using the dilution that gave 5000–10,000 Km(R) transductants per plate. Let sit 1 hour at room temperature to allow for phenotypic expression of Km(R). Plate 0.2 ml onto each of 20 LB plates with added Km (50 μg/ml). Incubate overnight at 37°.

3b. Since MudK insertions have to be in the correct orientation and reading frame, more colonies will have to be screened to obtain LacZ protein fusions (translational fusions) to genes of interest. Start a fresh 6 ml overnight culture of the *Salmonella enterica* recipient strain. For MudK insertions that require a screening step by replica printing add 5.5 ml of cells to 5.5 ml of diluted phage using the dilution that gave 5000–10,000 Km(R) transductants per plate. Let sit 1 hour at room temperature to allow for phenotypic expression of Km(R). Plate 0.2 ml onto each of 50 LB plates with added Km (50 μg/ml). Incubate overnight at 37°.

4. Add 0.3 ml of L broth to each plate and with a glass spreader suspend the Km(R) colonies in the L broth and transfer to a sterile flask. Collect and pool the cells together. Use 10 plates of cells to make each pool. Dilute cells from each pool into 20 ml of L broth to approximate the density of an overnight culture ($2 \times 10(9)$/ml). Pellet cells (8000 rpm in an SS34 rotor for 5 min) and resuspend in 20 ml of fresh L medium. Again pellet cells (8000 rpm in an SS34 rotor for 5 min) and resuspend in 20 ml of fresh L medium.

5. Add 80 ml of P22 broth to each washed 20 ml L broth pool and grow at 37° with aeration for 6 hours to overnight at 37°. Pellet cells (8000 rpm in an SS34 rotor for 5 min) and decant supernatant into a sterile flask. Add 2 ml $CHCl_3$ and shake to kill any remaining bacteria. Store at 4°. These pools will last for decades if kept in an sealed bottle. Each lysate represents 50,000 to 100,000 independent Mud insertion mutants.

6. Start a 2 ml overnight culture of a *Salmonella enterica* strain. Grow overnight with aeration at 37°.

7. Do a test cross with each pool lysate to determine the number of MudJ or MudK insertions obtained per 0.1 ml of diluted phage stock. Dilute the MudJ and MudK donor lysates 10^{-1}, 10^{-2}, and 10^{-3}. In a sterile tube, mix 0.1 ml of cells from the overnight culture with 0.1 ml of phage grown on the MudJ or MudK pools. A dilution that yields 5000–10,000 colonies per plate is a good working number.

8. Start a fresh 3 ml overnight culture of the *Salmonella enterica* recipient strain. Grow overnight with aeration at 37°. Add 2.2 ml of cells to 2.2 ml of diluted phage using the dilution that gave 5000–10,000 Km(R) transductants per plate. Plate 0.2 ml onto each of 20 LB plates with added Tc (15 μg/ml). Incubate overnight at 37°. Phenotypic expression of Tc resistance is not necessary.

9. Replica print L-Tc plates to L-Tc plates with added Km (50 μg/ml) and 1 mM EGTA to inhibit P22 growth in the colonies (Maloy *et al.*, 1996). Pick the Tc(R) Km(R) colonies and isolate P22-sensitive mutants using green indicator plates (Maloy *et al.*, 1996). These represent potential MudJ/K insertions in the gene of interest linked to the Tn10dTc marker. Screen for insertion in the gene of interest by PCR. Screen on lactose indicator media for Lac phenotypes.

Procedure 5: Conversion of MudJ/K insertions to MudA/B insertions by cassette replacement.

Materials

Strains with MudJ or MudK insertions in genes of interest.

P22 HT/*int* transducing lysates on *S. enterica* strains carrying MudA or MudB insertions in F plasmid DNA for cassette replacement experiments (TH331 = F'128 *zzf-1066*::MudA/*proAB47 pyrB64* and TH1123 = F'152 *zzf-1093*::MudB/*nadA56*).

1. Start a 1 ml overnight culture of a *Salmonella enterica* strain with a MudJ or MudK insertion in a gene of interest. Grow overnight with aeration at 37°.
2. To 0.1 ml of cells add 0.1 ml of a 10^{-2} dilution (in L broth) of the MudA donor for a MudJ recipient or the MudB donor for the MudK recipient. Plate on L Ap (30 μg/ml) plates and incubate overnight at 37°.
3. Pick 4 colonies and isolate P22-sensitive mutants using green indicator plates (Maloy *et al.*, 1996). Keep one colony that is Ap(R) Km(S). These will result from a conversion of the MudJ to MudA or MudK to MudB.

Procedure 6: Generation of tandem chromosomal duplication with a *lac* transcriptional fusion (MudA) or a *lacZ* translational fusion (MudB) to the gene of interest at the duplication join-point.

Materials

P22 HT/*int* transducing lysates prepared on MudA/B insertions in the gene of interest and on a MudA insertion in an auxotrophic gene that is UPSTREAM (5') of the target gene's promoter inserted in the same orientation on the chromosome of the MudA/B insertion in the target gene. Pick an auxotrophic marker that is close to the target gene (within 500 kbp, 10 min of the genome). NOTE: For an auxotrophic gene that is transcribed in the same direction on the chromosome as the target gene of interest (clockwise or counterclockwise based on the standard linkage map), a Lac$^+$ MudA insertion is in the same orientation in the chromo-some. For an auxotrophic gene that is transcribed in the opposite direction on the chromosome as the target gene of interest (clockwise or counter-clockwise based on the standard linkage map), a Lac$^-$ MudA insertion is in the same orientation in the chromosome.

1. Start a 1 ml overnight culture of a wild-type *Salmonella enterica* strain. Grow overnight with aeration at 37°.

2. Dilute the donor phage lysates 10^{-2} in L broth. Add 0.1 ml diluted phage to 0.1 ml cells directly on L Ap (30 μg/ml) plates and incubate overnight at 37°. Start a fresh 1 ml overnight culture of the wild-type *Salmonella enterica* strain. Grow overnight with aeration at 37°.

3. Using the number of Ap(R) transductants obtained from lysates grown on the MudA/B donor and the linked auxotrophic MudA strains determine phage dilutions for each lysate that will give equal numbers of Ap(R) transductants. Remember MudA and MudB are inherited by two-particle transduction events so a two-fold increase in the dilution will reduce the number of Ap(R) transductants by 4-fold. Add equal volumes of phage lysates diluted to produce equal numbers of Ap(R) transductants. Add 0.1 ml of the mixed lysate to 0.1 ml of cell culture onto L-Ap plates. Incubate overnight at 37°.

4. Replica print the L Ap plates to minimal salts glucose plates with added Ap (15 μg/ml) and L Ap plates with added Xgal.

5. Pick 12 prototrophic Ap(R) Lac$^+$ colonies and isolate P22-sensitive mutants using green indicator plates with added Ap (30 μg/ml) (Maloy *et al.*, 1996).

6. Streak each phage-sensitive isolate onto two L Xgal plates, one with added Ap (30 μg/ml) to hold the duplication and one without added Ap. Incubate overnight at 37°.

7. Any clone that gives a mixture of Lac$^+$ and Lac$^-$ colonies when streaked on media without Ap is a duplication recombinant with the *lac* transcriptional fusion for a MudA donor from the gene of interest or a *lacZ* translational fusion for a MudB donor from the gene of interest and an intact copy of the gene of interest.

8. If desired, the MudA or MudB insertion at the duplication join-point can be converted to a MudJ or MudK insertion using P22 lysates grown on either strain TH1380 (F'128 *zzf-1028*::MudJ/*proAB47 pyrB64*) to convert MudA to MudJ or strain TH3805 (F'152 *zzf-1093*::MudK/ *nadA56*) to convert a MudB to mudK. In both cases the selection is for Km(R) followed by a screen for Ap(S) and segregation of Lac$^-$ Km(S) colonies when grown in the absence of added Km to the medium. This will result in conversion of MudA/B at the duplication join-point to MudJ/K at the

duplication join-point. Then the duplication can be moved from strain to strain by a single particle P22 transduction selecting for Km(R) and screening for Lac segregation in the absence of added Km.

Procedure 7: Isolation of Tn*10* insertions linked to genes of interest for duplication construction. Complementation and dominance analysis.

Materials

P22 HT/*int* transducing lysates on *S. enterica* Tn*10* pools as described previously (Kleckner *et al.*, 1991).
P22 broth
Strain of *S. enterica* defective in the gene of interest.

1. Start a 1 ml overnight culture of the *Salmonella enterica* strain defective in the gene of interest. Grow overnight with aeration at 37°.

2. First do a test cross to determine the number of Tc(R) transductants obtained per 0.1 ml of diluted phage stock. Dilute the Tn*10* donor lysate 10^{-2}, 10^{-3}, and 10^{-4}. Plate 0.1 ml of cells from the overnight culture with 0.1 ml of phage grown on the Tn*10* pool donor directly onto L plates with added Tc (15 μg/ml). Start a fresh 3 ml overnight culture of the *Salmonella enterica* recipient strain defective in the gene of interest. Grow overnight with aeration at 37°.

3. Determine what dilution of the donor phage will give 300–500 Tc(R) colonies per plate. Add 2.2 ml of cells to 2.2 ml of diluted phage using the dilution that gave 300–500 Tc(R) transductants per plate. Plate 0.2 ml of the mixture onto each of 20 L Tc (15 μg/ml) plates. Incubate overnight at 37°.

4. Replica print onto Tc-containing media that will distinguish the wild-type and mutant allele in the gene of interest. For example, to isolate Tn*10* insertions linked to the *his* biosynthetic operon. A *his* auxotroph strain is used as the recipient. Each L-Tc plates is then replica printed to two minimal salts glucose Tc (7.5 μg/ml) plates, the first without added histidine and the second with added histidine. The plates are incubated overnight at 37° and the following day are screened for Tc(R) colonies that grow in media without added histidine. These will be Tn*10* insertion mutants that are linked to the *his* operon by P22 transduction.

5. Pick 20 colonies with Tn*10* insertions linked to the gene of interest and isolate P22-sensitive mutants using green indicator plates (Maloy *et al.*, 1996).

6. For each strain with a different linked transposon determine the co-transduction frequency between the Tn*10* insertion and the gene of interest. Grow a P22 transducing lysate on each of the 20 linked-Tn*10* insertion mutants (Procedure 1). Use each lysate to transduce the strain defective in the gene of interest to Tc(R).

7. Screen 100 Tc(R) transductants for those that have inherited the wild-type gene of interest. The number of wild-type Tc(R) recombinants is the percent cotransduction between the Tn*10* insertion and the gene of interest.

8. Determine the physical distance between the Tn*10* insertion and the gene of interest (Sanderson and Roth, 1988). Based on the approximate distance to the gene of interest design primers clockwise and counter-clockwise to the gene of interest to PCR amplify between these sites and a PCR primer to unique (non-IS*10)* Tn*10* sequence.

9. Once the location and orientation of the Tn*10* insertions are deter-mined, set up a cross with a donor strain with a selectable marker on one side of the gene of interest and a Tn*10* on the other side. Use this to transduce a recipient with a second Tn*10* on the side opposite the gene of interest and in the same orientation as the Tn*10* in the donor strain.

10. Transduce the recipient on media to select for the linked marker in the presence of Tc. If the donor strain has a wild-type allele for the gene of interest and the recipient has a defective recessive allele for the gene of interest, the duplication recombinants will be wild-type for the gene of interest. If both alleles are mutant, then the duplication recombinants will test for complementation of the mutant alleles. Pick 20 potential duplication recombinants and isolate P22-sensitive mutants using green indicator plates with added Tc (15 μg/ml) (Maloy *et al.*, 1996).

11. Streak out potential duplication recombinants on L medium without Tc. Incubate overnight at 37°.

12. Replica print to L and L-Tc plates and incubate overnight at 37°.

13. Score plates. Any strains that segregate Tc(S) colonies are duplica-tion recombinants. Score the Tc(S) segregants for the presence of alleles of the gene of interest used in the donor and recipient strains to verify that the duplication recombinant contained both donor and recipient alleles to validate and complement the dominance test.

References

Camacho, E. M., and Casadesus, J. (2001). Genetic mapping by duplication segregation in *Salmonella enterica*. *Genetics* **157,** 491–502.

Casadaban, M. J., and Chou, J. (1984). *In vivo* formation of gene fusions encoding hybrid beta-galactosidase proteins in one step with a transposable Mu-*lac* transducing phage. *Proc. Natl. Acad. Sci. USA* **81,** 535–539.

Casadaban, M. J., and Cohen, S. N. (1979). Lactose genes fused to exogenous promoters in one step using a Mu-*lac* bacteriophage: *In vivo* probe for transcriptional control sequences. *Proc. Natl. Acad. Sci. USA.* **76,** 4530–4533.

Chumley, F. G., Menzel, R., and Roth, J. R. (1978). Hfr formation directed by Tn*10*. *Genetics* **91,** 639–655.

Chumley, F. G., and Roth, J. R. (1980). Rearrangement of the bacterial chromosome using Tn10 as a region of homology. *Genetics* **94,** 1–14.

Groisman, E. A. (1991). *In vivo* genetic engineering with bacteriophage Mu. *Methods Enzymol.* **204,** 180–212.

Hughes, K. T., Ladika, D., Roth, J. R., and Olivera, B. M. (1983). An indispensable gene for NAD biosynthesis in *Salmonella typhimurium. J. Bacteriol.* **155,** 213–221.

Hughes, K. T., and Roth, J. R. (1984). Conditionally transposition-defective derivative of Mudl (Amp Lac). *J. Bacteriol.* **159,** 130–137.

Hughes, K. T., and Roth, J. R. (1985). Directed formation of deletions and duplications using Mud(Ap, *lac*). *Genetics* **109,** 263–282.

Hughes, K. T., and Roth, J. R. (1988). Transitory cis complementation: A method for providing transposition functions to defective transposons. *Genetics* **119,** 9–12.

Kleckner, N. (1983). Transposon Tn*10. In* "Mobile Genetic Elements" (J. A. Shapiro, ed.), pp. 261–298. Academic Press, New York.

Kleckner, N., Bender, J., and Gottesman, S. (1991). Uses of transposons with emphasis on Tn*10. Methods Enzymol.* **204,** 139–180.

Maloy, S. R., Stewart, V. J., and Taylor, R. K. (1996). "Genetic analysis of pathogenic bacteria." Cold Spring Harbor Press, Cold Spring Harbor, New York.

Pedulla, M. L., Ford, M. E., Karthikeyan, T., Houtz, J. M., Hendrix, R. W., Hatfull, G. F., Poteete, A. R., Gilcrease, E. B., Winn-Stapley, D. A., and Casjens, S. R. (2003). Corrected sequence of the bacteriophage p22 genome. *J.Bacteriol.* **185,** 1475–1477.

Sanderson, K. E., and Roth, J. R. (1988). Linkage map of *Salmonella typhimurium*, edition VII. *Microbiol. Rev.* **52,** 485–532.

[8] Target-Directed Proteolysis *In Vivo*

By Markus Eser, Tanja Henrichs, Dana Boyd, and
Michael Ehrmann

Abstract

The experimental problems associated with *in vivo* studies of essential proteins or integral membrane proteins have triggered geneticists to generate novel approaches that have often led to insights of general relevance (Shuman and Silhavy, 2003). In order to extend the experimental portfolio, we developed target-directed proteolysis (TDP), an *in vivo* method allowing structural and functional characterization of target proteins in living cells. TDP is based on the activity of the highly sequence-specific NIa protease from tobacco etch virus. When its recognition site of seven residues is engineered into target proteins and NIa protease is expressed under tight promoter control, substrates can be conditionally processed while other cellular proteins remain unaffected. Applications include conditional inactivation as well as functional characterization of target proteins.

METHODS IN ENZYMOLOGY, VOL. 421
Copyright 2007, Elsevier Inc. All rights reserved.

0076-6879/07 $35.00
DOI: 10.1016/S0076-6879(06)21008-5

Introduction

Functional characterization of proteins *in vivo* is often done by mutational analysis. The basic principle is to carry out either random or site-specific mutageneses followed by phenotypic characterization using direct or indirect reporter assays. Consequently, experimental procedures may be improved by establishing additional genetic or analytical techniques. This chapter describes a novel method that uses a biochemical activity as a tool in genetic experiments.

As most of the important and evolutionarily conserved metabolic and regulatory processes are carried out by cellular factors that are essential for viability, conditional inactivation of the relevant proteins presents an experimental challenge. The most common technique is depletion during growth following cessation of synthesis of new, active gene product. Depletion can be accomplished by a variety of means, including shut-off of the gene itself or of a nonsense suppressor of a nonsense mutant allele of the gene by a tightly regulated promoter or a temperature-sensitive allele of either the gene or the suppressor. Depletion of the gene product occurs gradually as a consequence of growth after cessation of synthesis. Most temperature-sensitive alleles have a defect in assembly of the gene product at the nonpermissive temperature but do not result in a thermolabile protein. Some temperature-sensitive alleles, however, do confer thermolability to the gene product. In such cases, loss of activity is much faster than that obtained by simple depletion since the gene product is actively inactivated by thermal denaturation. Such mutants are desirable since rapid loss of activity minimizes secondary effects, phenotypic changes that are only indirect consequences of the depletion of the gene product; however, such mutants are rare (Boyd *et al.*, 1968). We have extended the portfolio of approaches allowing rapid, controlled, and conditional inactivation by developing TDP that allows proteolytic inactivation of target proteins *in vivo*. TDP uses engineering of target proteins to contain a TEV NIa protease cleavage site and controlled co-expression of the protease. Polypeptides that are not natural substrates of TEV protease are proteolyzed if they carry the appropriate cleavage site. Because of its distinct specificity, other proteins are not affected by TEV protease that can be expressed in various cellular compartments without interfering with viability.

TEV Protease

The tobacco etch virus genome encodes one polyprotein of about 350 kDa that is subsequently proteolytically processed into more than a dozen individual proteins, three of which are proteases. The main processing factor is the small nuclear inclusion (NIa) protease, which is homologous to picornavirus 3C protease (Carrington and Dougherty, 1987a,b). Its C-terminal fragment of 27 kDa is sufficient for proteolytic activity (Carrington and Dougherty, 1987a,b). It is this fragment, termed TEV protease, that is used as a tool for various site-specific proteolysis approaches. The protease recognizes a seven amino acid consensus sequence, Glu-X-X-Tyr-X-Gln/Ser, where X can be various amino acyl residues (Dougherty *et al.*, 1989). Cleavage occurs between the conserved Gln and Ser residues (Carrington and Dougherty, 1988). TEV protease is a Cys protease. It is thus inhibited by thiol alkylating reagents such as iodoacetamide. Recently, its crystal structure has been determined, which should provide opportunities for engineering to improve its catalytic parameters or to change its specificity (Phan *et al.*, 2002). The simplest and therefore perhaps the most well-known application of TEV protease is the removal of affinity tags of recombinant proteins. However, this protease can also be used in various *in vivo* applications such as the inactivation of essential proteins, and the mapping of functions to specific domains, as well as genetic screening procedures.

TEV Protease Expression Vectors

In order to process cytoplasmic target proteins, various expression vectors were constructed. These vectors should be under tight promoter control, as preferentially no TEV protease should be produced under noninducing conditions. For this purpose, medium or low copy-number vectors, such as pBR322, pACYC184, or pHSG575, are recommended. To limit TEV protease levels under noninducing conditions, the *lac* or *tet* promoters can be used, particularly when the *lac* or *tet* repressor genes are present on the expression vectors. The best control of protease activity was obtained when using low copy-number plasmids expressing TEV protease under P*tet* control such as pTH9 (Henrichs *et al.*, 2005). It should be noted that TEV protease does not lose its activity if it is part of a hybrid protein, as the presence of glutathione-S-transferase and maltose-binding protein does not interfere with protease activity.

Before performing functional characterization, it is essential to determine optimal conditions, which might vary greatly depending on the experimental system. It is recommended that the best inactivation conditions be established by testing various inducer concentrations (e.g., 0.005 to 0.2 μg/ml anhydrotetracycline) at various growth phases (early and mid-log phase cultures) by following expression on Western blots using the available polyclonal antibodies (Faber *et al.*, 2001). Titrating the levels of target protein and monitoring the viability of cells (in simple growth tests) is also required to ensure conditions that are close to wild-type. It is obvious that functional characterization should not be performed with dead cells. It might also be helpful to test various temperatures and growth media, as sometimes the function of a target protein might be more or less required under some conditions. If precise knowledge of the kinetics of proteolytic processing is required, pulse chase experiments will provide this information.

Introducing a TEV Protease Recognition Site

Random Insertion via TnTIN and of TnTAP

Two Tn5-based minitransposons are available to insert TEV protease cleavage sites at random into target proteins (Fig. 1) (see Ehrmann *et al.*, 1997, for details). TnTIN introduces TEV protease cleavage sites into cytoplasmic proteins. TnTAP facilitates the same operation for proteins localized to the bacterial cell envelope. These transposons consist of 19 bp OE of Tn5 required for transposition, a 7-codon TEV protease cleavage site, *uidA* or signal sequenceless *phoA*, *neo*, and another 19 bp OE of Tn5. The Tn5 transposase, acting on the flanking OE sequences, is supplied *in trans*, to generate stable insertions. When inserted in the correct orientation and reading frame, the transposons form translational fusions of the TEV protease site and the reporter genes *neo* or *phoA* with the target gene. These translational fusions can be conveniently detected on indicator plates containing specific dyes, that is, 5-Bromo-4-chloro-3-indolyl β-D-glucuronide (X-Gluc) for *uidA* fusions, and 5-bromo-4-chloro-3-indolyl phosphate (X-P) for *phoA* fusions. Both transposons carry *NotI* restriction sites flanking the codons for the reporter genes and *neo*. Thus, *NotI* deletions remove *uidA* or *phoA* and *neo*, leaving a 72-bp insert in an otherwise intact target gene. The insertion generated by the transposons is LTLIHKF<u>ENLYFQ/ S</u>AAAILVYKSQ. The TEV protease recognition sequence is ENLYFQ/S.

Via pEDIE2/3

For site-directed insertions of TEV protease cleavage sites, plasmid-based systems are available. pEDIE2 and 3 allow cloning of PCR fragments of the target gene, and thus precise placement of inserts (Harnasch *et al.*, 2004). This strategy is used when high-resolution structural and functional information for the target protein is already available. These plasmids have features similar to TnTIN and TnTAP that facilitate detection of positive clones (Fig. 2). In addition, the tightly controlled arabinose promoter allows simultaneous depletion in combination with TDP, a strategy that might improve conditional inactivation.

Applications of TDP

Inactivation of Essential Proteins in the Cytoplasm

During post-translational secretion, the widely conserved and essential SecA protein recognizes secretory proteins and carries them to the

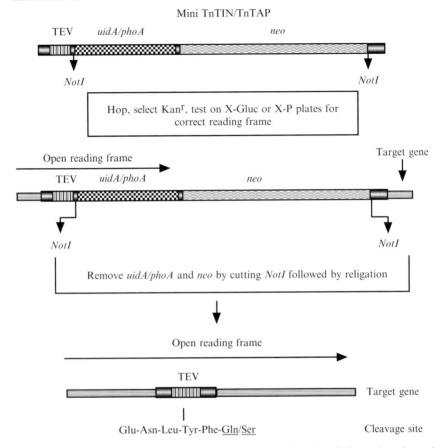

Fɪɢ. 1. Transposons for insertion of TEV protease sites. To isolate in-frame insertions of TnTIN or TnTAP into target genes, these minitransposons are transformed into cells containing a plasmid expressing the target genes. During growth overnight, transposition occurs. Plasmids containing transposon insertions are digested with *NotI* and re-ligated, deleting *uidA* (*phoA*) and *neo*. The remaining insert of 72 bp encodes LTLIHKFENLYFQ/SAAAILVYKSQ. TEV protease recognition sequence is ENLYFQ/S. The resulting plasmids can be verified by restriction analysis and expression of target genes containing the desired insert can be detected on Western blots using antibodies against the target proteins since these TEV protease site derivatives migrate at slightly higher MW on SDS-PAGE.

translocon. While interacting with the translocon, SecA energizes secretion via ATP hydrolysis (Mori and Ito, 2001). During targeting of soluble secretory proteins, SecA performs a quality control function that is based on a general chaperone activity. This quality control mechanism involves assisted folding of signal sequenceless proteins, thereby excluding them from the secretion process (Eser and Ehrmann, 2003). In contrast to most

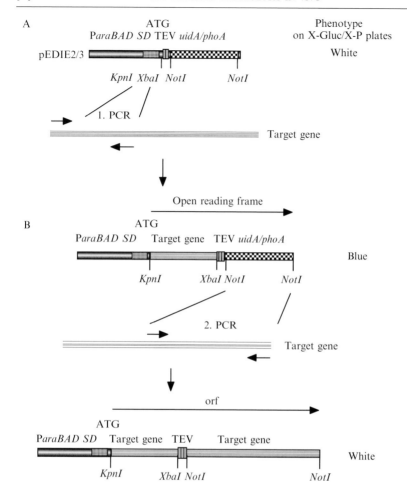

Remaining insert after cloning into *KpnI-XbaI* and *NotI* of pEDIE2

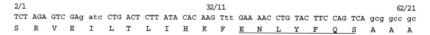

```
2/1                              32/11                              62/21
TCT AGA GTC GAg atc CTG ACT CTT ATA CAC AAG Ttt GAA AAC CTG TAC TTC CAG TCA gcg gcc gc
 S   R   V   E   I   L   T   L   I   H   K   F   E   N   L   Y   F   Q   S   A   A   A
```

FIG. 2. Plasmid system to generate TEV protease sites. pEDIE2 and 3 allow cloning of PCR products in front and after the TEV protease site. The introduction of the 5′ fragment of the target gene generates an in-frame fusion of the target gene, the TEV protease site and the reporter gene *uidA*, encoding β-glucuronidase, or *phoA*, encoding alkaline phosphatase. The activity of the reporter genes facilitates the isolation of successful clones. In a second cloning step, a PCR fragment of the remaining 3′ fragment of the target gene is inserted leading to a replacement of the reporter genes. In addition, pEDIE 2 and 3 express the target gene under control of the tightly controlled arabinose promoter P*araBAD*.

soluble proteins, targeting of integral membrane proteins is co-translational, involving the components of the signal recognition particle (SRP) system. This includes the SRP itself, composed of the Ffh protein and the 4.5S RNA, and the SRP-receptor FtsY (reviewed in Herskovits *et al.*, 2000). In some cases the additional requirement of SecA in membrane insertion has been demonstrated but it is believed that in this case SecA is not involved in targeting but in the translocation of periplasmic domains of the membrane proteins (for review see Driessen *et al.*, 2001). To analyze the contribution of SecA to membrane insertion, conditional inactivation and convenient membrane insertion assays are required. To establish TDP of SecA three inserts of the TEV protease recognition site were identified that in the absence of TEV protease did not interfere with SecA function. Here, TEV protease sites were engineered after codons 195, 252, and 830 and chromosomal wt *secA* was replaced with these *secA* constructs (Ehrmann *et al.*, 1997; Mondigler and Ehrmann, 1996).

Proteolytic Inactivation of SecA

While SecA 252 and 830 are efficiently proteolyzed by co-expressed TEV protease *in vivo* and *in vitro*, no functional defects were detected (Ehrmann *et al.*, 1997; Mondigler and Ehrmann, 1996). These data indicate that SecA252 and 830 represent truly permissive sites that tolerate the insertion of additional residues as well as proteolytic processing. In contrast, SecA195 is only partially processed but this event leads to a weak secretion defect (Mondigler and Ehrmann, 1996). This result is typical for a nonoptimally positioned cleavage site. In general, cleavage sites must be surface exposed in order to be accessible for the protease. In addition, the TEV protease recognition site must be flexible to adopt a conformation allowing productive interaction with the active site for efficient processing. The obvious experimental adjustment to this situation is to add a linker (e.g., Ser-Gly$_3$) before and after the cleavage site, which has often improved the efficiency of processing. An alternative and more global solution is to alter the experimental conditions such that the substrate is cleaved before it is completely folded.

Proteolysis at the Ribosome

Recent crystallographic studies of the ribosome and its accessory proteins that are lining the exit channel for nascent chains show that the proline isomerase trigger factor (Tig) attaches via ribosomal protein L23 to the 50S subunit (Ferbitz *et al.*, 2004). It is the N-terminal domain of Tig that is responsible for ribosomal localization while its C-terminal PPIase domain is not. Thus, when TEV protease is tethered to the N-terminus of

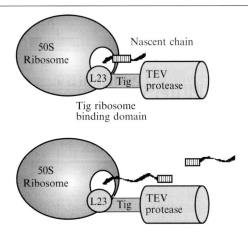

FIG. 3. TDP at the ribosome. TEV protease is fused to the ribosome-binding domain of Tig interacting with ribosomal protein L23, thus lining the exit channel for nascent chains. When a nascent chain emerges from the exit channel of the ribosome containing a TEV protease cleavage site, TDP will occur before the target protein is completely folded.

Tig, it is positioned closely to the exit channel of the ribosome (Fig. 3). Due to the antagonistic activities of protein translation and proteolytic degradation, fusing a protease to the ribosome does not appear to be feasible. However, due to its high specificity, TEV protease is well tolerated and does not result in detectable toxicity (Henrichs *et al.*, 2005). In addition, Tig is a nonessential protein in *Escherichia coli* that can be deleted without loss of viability. Another advantage might be that Tig is not exclusively localized to the ribosome, but significant amounts are also present in the cytoplasm. This nonexclusive localization might be beneficial because cytoplasmic TEV protease can move around and cleave substrates that have either escaped processing during translation or have been synthesized before induction of TEV protease.

Ribosomal localization of TEV protease leads to efficient processing of SecA195, resulting in secretion defects and a reduction of cell growth in rich media, which in turn indicates a strong improvement of the method. When comparing two approaches of SecA inactivation, depletion and TDP, we detected a stronger and more rapid effect on cell growth under depletion conditions compared to TDP inactivation. This result might suggest that proteolytic inactivation of SecA195 is not as efficient compared to SecA depletion. However, similar translocation defects were observed following depletion and proteolytic inactivation when using the soluble alkaline phosphatase or the cytoplasmic membrane proteins FtsQ and MalF as reporters, suggesting that TDP produces significant secretion phenotypes (Henrichs *et al.*, 2005). It should also be considered that working with healthier strains is expected to yield more direct phenotypes compared to cells that have completely stopped dividing.

While this system allowed us to test the implications of the essential targeting factor SecA in co-translational membrane insertion, it also improved the method of target-directed inactivation, but perhaps most importantly provides the first experimental evidence that novel biological activities can be recruited to ribosomes without interfering with viability. Such biological activities of interest might include any type of post-translational modification or molecular chaperones. It can also be considered to attach less specific proteases to the ribosome. These proteases could, for example, readily degrade proteins that misfold during co-translational folding. It might be advantageous to remove such proteins from ribosomes early to prevent entry into the aggregation pathway, a strategy that might lead to healthier strains and thus to optimized production of recombinant proteins.

Structure–Function Studies of Integral Outer Membrane Proteins

After insertion of TEV protease sites into an outer membrane protein, surface accessibility of cleavage sites can be tested by adding purified TEV protease to whole cells. Theoretically, TEV protease sites located on the cell surface are processed while periplasmic sites are not because TEV protease cannot cross the outer membrane. Therefore, treatment of whole cells should identify segments of the target outer membrane protein that are exposed to the cell surface. A number of outer membrane proteins have been subjected to TDP. For some proteins such as TolC and LamB, crystal structures are available (Koronakis *et al.*, 2000; Schirmer *et al.*, 1995), and the obtained proteolysis data corresponded with the structural information (T. Henrichs *et al.*, manuscript in preparation). For other proteins, for which high-resolution structural information is not available, such as PulD (Guilvout *et al.*, 1999), FhaC (Guedin *et al.*, 2000), and PapC (Henderson *et al.*, 2004), the data were not as clear-cut. For example, in PulD, TDP, and site-directed fluorescence labeling, data were sometimes inconsistent, suggesting that more work is required to optimize the experimental procedures for the determination of outer membrane protein topology.

As the insertion of TEV protease sites also represents a linker mutagenesis experiment, phenotypic characterization of these mutants can provide functional information, and if reliable structural information is available, particular functions can be assigned to relevant domains. For example, low expression levels of mutants containing TEV protease recognition sequences can indicate assembly defects. In addition, as many outer membrane proteins form SDS-resistant but heat-labile trimers, trimer formation can be monitored by comparing the migration of the insertion mutants on SDS-PAGE after subjecting (or not subjecting) samples to boiling for 30 min before loading on SDS-PAGE (Guedin *et al.*, 2000; Guilvout *et al.*, 1999). In the case of TolC (Koronakis *et al.*, 2004) and LamB (Boos and Shuman, 1998), where the target protein has well-defined functions that can be conveniently assayed, the insertion mutants can provide useful information. Such phenotypic analyses, including, for example, uptake or secretion of substrate molecules or the sensitivity to phage infections, allow the mapping of specific functions to discrete structural regions.

Standard TEV Protease Assay of Whole Cells

Proteins exposing a TEV protease site at the cell surface can be proteolyzed by adding TEV protease to whole cells. A typical assay procedure is:

1. Grow cells to OD_{600} = 0.3–0.5.
2. Spin and wash in TEV protease buffer.
3. Resuspend to OD = 0.3.

Protease Assay

1. Place 180 μl cells in a protease buffer (OD = 0.3) into an Eppendorf tube.
2. Add 2 μl protease buffer 10X.
3. Add 0.2 μl DTT (100 mM stock).
4. Add 1 to 5 μl protease (1 μg/μl).
5. Incubate 1 to 4 h at 30°.
6. Add 100 mM iodoacetamide to stop reaction.

Sample Analysis

1. Resuspend pellet in 50-μl sample buffer.
2. Boil 10 min.
3. Run half of the sample on SDS PAGE/Western blot.

Protease Buffer 10X

   0.5 M of Tris HCl, pH = 8.0
   5 mM EDTA (ethylenediaminetetraacedic acid)

TCA Precipitation

1. Add 190 μl TCA (20% stock).
2. Vortex and incubate on ice 15 min.
3. Centrifuge for 20 min at 4°.
4. Remove supernatant.

5. Wash once with ice-cold acetone.
6. Centrifuge for 20 min at 4°.
7. Remove supernatant.
8. Dry pellet under vacuum.
9. Resuspend pellet in 20 μl sample buffer.

Purification of TEV Protease

Various standard vector systems producing high levels of TEV protease such as T7 polymerase vectors can be used. Expression while growing cells at 20° will lead to mainly soluble TEV protease, while growth at 37° will produce inclusion bodies. Alternatively, inclusion body formation can be utilized as a purification step since TEV protease can be efficiently refolded. N-terminal (His)$_6$ tags facilitate purification and do not interfere with activity or refolding. Yields of up to 30 mg of purified TEV protease/l of culture can be expected.

Purification of Soluble 6His TEV Protease

1. Grow cells in double-rich medium at 20° to OD = 0.8. (If TEV protease expression is IPTG dependent, do not use BactoTryptone in the medium because it contains lactose. Change BactoTryptone by NZ amine A (available from Sigma).
2. Induce (e.g., 0.01 to 0.1 mM IPTG).
3. Let grow overnight.
4. Harvest and wash cells (e.g., in 50 mM NaH$_2$PO$_4$, 100 mM NaCl, pH 8.0).
5. Lyse cells (e.g., by passing twice through a French press [20,000 psi]).
6. Purify TEV protease by Ni-chromatography using standard protocol supplied by, for example, Qiagen.
7. After elution with imidazole, dialyze immediately against 50 mM of Tris HCl pH=7.5 plus 1 mM of EDTA, and 5 mM of DTT.
8. Do not exceed concentration of TEV protease (not over 0.5 to 1 mg/ml).
9. Spin 20,000 × g for 30 min.
10. Take supernatant.
11. Add 10% glycerol (final concentration) and store aliquots at −80°. Freezing in the absence of glycerol will abolish activity.

Purification of Insoluble (His)$_6$-TEV Protease

1. Grow cells in double-rich medium at 37° to OD$_{600}$ = 0.8.
2. Fully induce (e.g., 1 mM of IPTG).

3. Grow overnight at 37°.
4. Harvest, wash cells; for example, use 50 mM of NaH$_2$PO$_4$, and 100 mM of NaCl, pH 8.0.
5. Lyse cells; for example, passing twice through a French press (20,000 psi).
6. Harvest inclusion bodies by centrifugation (27,000 × g, 20 min, 4°).
7. Solubilize inclusion bodies by resuspending the pellet in 8 M of urea, 0.1 M of NaH$_2$PO$_4$, and 0.01 M of Tris/HCl, pH 8.0.
8. Remove insoluble aggregates by centrifugation (27,000 × g, 10 min, room temperature).
9. Purify TEV protease by Ni-chromatography using standard protocol.
10. After elution with imidazole, dialyze against 8 M of urea in 50 mM of Tris HCl, pH = 7.5, plus 1 mM of EDTA (ethylenediaminetetraacetic acid) and 5 mM of DTT (dithiothreitol).
11. Refold TEV protease by dilution.

TEV Protease Storage Buffer

50 mM of Tris HCl, pH = 7.5.
1 mM of EDTA.
5 mM of DTT.
50% glycerol.
0.1% Triton X100.
Store at −80°.

Additional Practical Considerations

As with any other method, TDP has limitations, some of which are generally relevant. For example, this method requires that a cleavage site is surface exposed. Therefore, best results will be obtained when a TEV protease recognition site is placed between two proteins or separate domains. If this is not possible, analysis of secondary structure and surface accessibility using bioinformatics will be helpful. In line with this rule, short loops connecting two transmembrane segments are often not cleaved. Equally obvious is that insertion of the cleavage site should not interfere with protein function. If a three-dimensional structure or detailed structure/function data are unavailable, an amino acid sequence alignment of the members of the same protein family combined with secondary structure predictions helps to identify likely candidate sites. Another general rule is to avoid overexpression of target proteins, which is rarely a good idea for any biological experiment.

The activity of TEV protease is weaker compared to classical proteases such as trypsin or proteinase K. Therefore, higher TEV protease concentrations might sometimes be required to obtain similar results.

Spheroplast buffers and detergents inhibit TEV protease. The protease is also sensitive to 0.01% SDS, but tolerates 0.01% TritonX100 and 0.01% dodecyl maltoside.

Overexposure of Western blots can identify cleavage sites that are not efficiently processed. Such results can lead to incorrect models. Therefore, carrying out a number of titration experiments is recommended, by varying, for example, the expression levels of the target protein, the assay temperature (keeping in mind that protein structures may change as a result of the temperature), and cell density (when performing TDP on whole cells, lower cell densities have yielded better results).

TEV protease is sometimes unstable even when expressed at a reasonable level. For example, when we expressed TEV protease in the periplasm, proteolytic activity was present even though TEV protease was not detectable under steady-state conditions when monitored by Western blotting.

Additional Potential Applications

TEV protease has the potential to develop into a more generally applicable tool, and it is up to the imagination of the reader to think of additional applications. One obvious scenario in which the use of TEV protease could lead to interesting results includes genetic screening or selection methods. In fact, TDP was first developed when we devised a genetic *in vivo* selection for *E. coli* mutants defective in membrane protein insertion (Fig. 4). The system is based on a reporter membrane protein composed of the N-terminal two transmembrane segments of MalF fused to alkaline phosphatase. This MalF-PhoA fusion was engineered to contain a TEV protease cleavage site in the periplasmic domain followed by an additional Arg residue. In wild-type cells, membrane insertion would be rapid and cytoplasmic TEV protease is unable to cleave the reporter protein. In this situation, alkaline phosphatase is inactive, because it requires periplasmic localization for proper folding. When membrane insertion is slowed down, TEV protease will cleave the future periplasmic domain in the cytoplasm. Subsequently, the C-terminal fragment containing the second transmembrane segment and alkaline phosphatase will insert into the membrane in the opposite orientation compared to the full-length reporter protein. This is because the engineered Arg residue is the only orientational determinant of this C-terminal membrane protein fragment. As alkaline phosphatase will be secreted and active, growth on β-glycerol phosphate agar plates can be used to specifically select for mutants defective in membrane insertion.

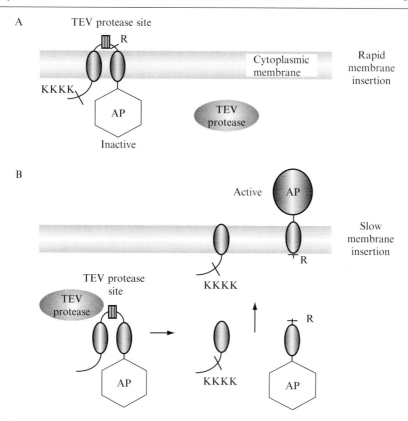

FIG. 4. A genetic selection for membrane insertion mutants. Cytoplasmically expressed TEV protease will cleave the reporter membrane protein MalF-AP, causing periplasmic localization of AP. The activity of periplasmic AP can be used to select for the desired mutants on appropriate media (see text for details).

Acknowledgments

This work has been funded by Biotechnology and Biological Sciences Research Council (BBSRC) and German Israeli Foundation (GIF [M.E.]). We would like to thank Hiroshi Kadokura, Eitan Bibi, Klaas Nico Faber, Tony Pugsley, and Vandana Gupta for discussions.

References

Boos, W., and Shuman, H. (1998). Maltose/maltodextrin system of *Escherichia coli*: Transport, metabolism, and regulation. *Microbiol. Mol. Biol. Rev.* **62,** 204–229.
Boyd, D., Nixon, R., Gillespie, S., and Gillespie, D. (1968). Screening of *Escherichia coli* temperature-sensitive mutants by pretreatment with glucose starvation. *J. Bacteriol.* **95,** 1040–1050.

Carrington, J. C., and Dougherty, W. G. (1987a). Small nuclear inclusion protein encoded by a plant potyvirus genome is a protease. *J. Virol.* **61,** 2540–2548.

Carrington, J. C., and Dougherty, W. G. (1987b). Processing of the tobacco etch virus 49K protease requires autoproteolysis. *Virology* **160,** 355–362.

Carrington, J. C., and Dougherty, W. G. (1988). A viral cleavage site cassette: Identification of amino acid sequences required for tobacco etch virus polyprotein processing. *Proc. Natl. Acad. Sci. USA* **85,** 3391–3395.

Dougherty, W. G., Cary, S. M., and Parks, T. D. (1989). Molecular genetic analysis of a plant virus polyprotein cleavage site: A model. *Virology* **171,** 356–364.

Driessen, A. J., Manting, E. H., and van der Does, C. (2001). The structural basis of protein targeting and translocation in bacteria. *Nat. Struct. Biol.* **8,** 492–498.

Ehrmann, M., Bolek, P., Mondigler, M., Boyd, D., and Lange, R. (1997). TnTIN and TnTAP: Mini-transposons for site specific proteolysis *in vivo*. *Proc. Natl. Acad. Sci. USA* **94,** 12111–13115.

Eser, M., and Ehrmann, M. (2003). SecA-dependent quality control of intracellular protein localization. *Proc. Natl. Acad. Sci. USA* **100,** 13231–13234.

Faber, K. N., Kram, A. M., Ehrmann, M., and Veenhuis, M. (2001). A novel method to determine the topology of peroxisomal membrane proteins *in vivo* using the tobacco etch virus protease. *J. Biol. Chem.* **276,** 36501–36507.

Ferbitz, L., Maier, T., Patzelt, H., Bukau, B., Deuerling, E., and Ban, N. (2004). Trigger factor in complex with the ribosome forms a molecular cradle for nascent proteins. *Nature* **431,** 590–596.

Guedin, S., Willery, E., Tommassen, J., Fort, E., Drobecq, H., Locht, C., and Jacob-Dubuisson, F. (2000). Novel topological features of FhaC, the outer membrane transporter involved in the secretion of the *Bordetella pertussis* filamentous hemagglutinin. *J. Biol. Chem.* **275,** 30202–30210.

Guilvout, I., Hardie, K., Sauvonnet, N., and Pugsley, A. (1999). Genetic dissection of the outer membrane secretin PulD: Are there distinct domains for multimerization and secretion specificity? *J. Bacteriol.* **181,** 7212–7220.

Harnasch, M., Grau, S., Dove, S., Hochschild, N., Iskandar, M-K., Xia, W., and Ehrmann, M. (2004). Bacterial expression and two-hybrid systems for human membrane proteins: Characterisation of presenilin/amyloid precursor interaction. *Mol. Membrane Biol.* **21,** 373–383.

Henderson, N. S., Shu Kin So, S., Martin, C., Kulkarni, R., and Thanassi, D. G. (2004). Topology of the outer membrane usher PapC determined by site-directed fluorescence labeling. *J. Biol. Chem.* In press.

Henrichs, T., Mikhaleva, N., Conz, C., Deuerling, E., Boyd, D., Zelazny, A., Bibi, E., Ban, N., and Ehrmann, M. (2005). Target-directed proteolysis at the ribosome. *Proc. Natl. Acad. Sci. USA* **102,** 4246–4251.

Herskovits, A. A., Bochkareva, E. S., and Bibi, E. (2000). New prospects in studying the bacterial signal recognition particle pathway. *Mol. Microbiol.* **38,** 927–939.

Koronakis, V., Sharff, A., Koronakis, E., Luisi, B., and Hughes, C. (2000). Crystal structure of the bacterial membrane protein TolC central to multidrug efflux and protein export. *Nature* **405,** 914–919.

Koronakis, V., Eswaran, J., and Hughes, C. (2004). Structure and function of TolC: The bacterial exit duct for proteins and drugs. *Annu. Rev. Biochem.* **73,** 467–489.

Mondigler, M., and Ehrmann, M. (1996). Site specific proteolysis of the *Escherichia coli* SecA protein *in vivo*. *J. Bacteriol.* **178.,** 2986–2988.

Mori, H., and Ito, K. (2001). The Sec protein-translocation pathway. *Trends Microbiol.* **9,** 494–500.

Phan, J., Zdanov, A., Evdokimov, A. G., Tropea, J. E., Peters, H. K., 3rd, Kapust, R. B., Li, M., Wlodawer, A., and Waugh, D. S. (2002). Structural basis for the substrate specificity of tobacco etch virus protease. *J. Biol. Chem.* **277**, 50564–50572.

Schirmer, T., Keller, T., Wang, Y., and Rosenbusch, J. (1995). Structural basis for sugar translocation through maltoporin channels at 3.1 A resolution. *Science* **267**, 512–514.

Shuman, H. A., and Silhavy, T. J. (2003). The art and design of genetic screens: *Escherichia coli*. *Nat. Rev. Genet.* **4**, 419–431.

[9] Sets of Transposon-Generated Sequence-Tagged Mutants for Structure–Function Analysis and Engineering

By BETH TRAXLER and ELIORA GACHELET

Abstract

Various genetic strategies are available for the isolation of small, in-frame insertional mutants. Here, we summarize some of the ways in which the resulting mutant libraries in particular genes have been used for the analysis of protein structure–function relationships and in engineering applications.

Introduction

Various genetic strategies are available to isolate libraries of mutants containing small in-frame sequence tags for proteins in bacteria (see reviews by Manoil and Traxler, 2000; Manoil, 2000; Gallagher *et al.*, 2006). These strategies exploit the ease of generating transposon insertions *in vitro* or *in vivo*. Depending on the method used, one might isolate numerous different insertions into a gene of interest and then exploit those mutants to study topics as diverse as gene regulation, the role of the gene product for particular pathways, or protein structure and folding. Suitable transposon insertions are usually identified initially via the expression of a translational fusion protein such as β-galactosidase or alkaline phosphatase (LacZ or PhoA). A subsequent processing step removes the majority of the transposon sequences but leaves behind a scar at the original site of the insertion, resulting in additional residues inserted into the polypeptide during translation. The size of the insertional scar (or sequence tag) varies, depending on the mutagenesis method, but usually ranges between 24 and 63 codons (Ehrmann *et al.*, 1997; Gallagher *et al.*, 2006; Manoil and Bailey, 1997). Useful features such as an antigenic epitope or a protease cleavage site are incorporated into the sequence tag, which enable specific detection or manipulation of the tagged mutant proteins.

METHODS IN ENZYMOLOGY, VOL. 421
Copyright 2007, Elsevier Inc. All rights reserved.

0076-6879/07 $35.00
DOI: 10.1016/S0076-6879(06)21009-7

The work of several labs (including our own) has demonstrated that libraries of insertion-tagged mutants can provide a powerful tool for the analysis of a variety of proteins, including soluble cytoplasmic or extracyto-plasmic proteins and integral membrane proteins, whose origins range from *Escherichia coli* and *Salmonella* to other prokaryotic and eukaryotic organisms. An earlier review of this methodology summarized several uses of these sequence-tagged mutants (Manoil and Traxler, 2000); here, we will re-emphasize a few of those previously discussed points and review more recent studies and applications.

Permissive Sites

We will focus primarily on mutant protein sets generated via Tn*lacZ*/in or Tn*phoA*/in transposon mutagenesis. The original description of these transposon tools detailed their use as a strategy to identify "permissive sites" or surface-exposed regions of proteins, which could tolerate the insertion of 31 additional residues without compromising the activity of the protein (Manoil and Bailey, 1997). The range of permissive sites identified within a set of "i31" mutants varies broadly: frequently, 20 to 25% of the i31 mutants in a particular protein have essentially normal activity, but several exceptions to this are noted in Table I, which represents a partial list of proteins characterized using transposon-generated small insertional tags. The permissive site mutants represent a valuable resource, especially for proteins without an available high-resolution structure (where the mutants provide indications of aqueous-exposed regions). In addition, the insertions at permissive sites can provide an antigenic epitope, allowing the detection of the tagged functional protein or a way to create a novel conditional protease-sensitive mutant (e.g., Ehrmann *et al.*, 1997; Kennedy and Traxler, 1999; Lee *et al.*, 1999).

Beyond Permissive Sites

In general, many i31 insertional mutations express proteins that are stable in the cell, even if the mutant proteins have lost their normal activity. (The broad exception to this is if the i31 insertion occurs in the region of a gene that codes for a transmembrane domain. The i31 sequence is highly charged and is usually incompatible with the hydrophobic membrane environment. Such mutant proteins are rarely stable after synthesis. Examples of this include mutants identified by Manoil and Bailey, 1997; Nelson *et al.*, 1997; Lui *et al.*, 2006) The stability of many non-functional i31 proteins allows the researcher to investigate the loss of function phenotype, given the availability of suitable tests. One example of this was the analysis of the

TABLE I

REPRESENTATIVE PROTEINS CHARACTERIZED WITH IN-FRAME INSERTIONAL MUTANTS

Protein/Source[a]	Insertion Size/Tn	Permissive[b]/Total Mutants	References
LacY/E. coli IM	31 aa; TnlacZ/in	9/21	Manoil and Bailey, 1997
LacI/E. coli cyto	31 aa; TnlacZ/in	8/18	Nelson et al., 1997
SecA/E. coli cyto and IM	24 aa; TnTIN	At least 1/2[c]	Ehrmann et al., 1997
TolC/E. coli OM	24 aa; TnTAP	At least 4/5[c]	Ehrmann et al., 1997
MalK/E. coli cyto and IM	31 aa; TnlacZ/in	7/13	Lippincott and Traxler, 1997
MalG/E. coli IM	31 aa; TnlacZ/in or TnphoA/in	1/18	Nelson and Traxler, 1998
MalF/E. coli IM	31–51 aa; TnlacZ/in, TnphoA/in, ISΩ/hah	8/42 (29 i31, 13 i51)	Gachelet, Talic, et al., in preparation
Mtv-7 Sag/Mouse mammary tumor virus in B cell PM	31 aa; TnlacZ/in	1/14	McMahon et al., 1998
TraD/E. coli IM	31 aa; TnlacZ/in or TnphoA/in	3/9	Lee et al., 1999
pCF10 AS (PrgB)/E. faecalis cell surface	31 aa; TnlacZ/in or TnphoA/in	10/23	Waters and Dunny, 2001
FliC/Salmonella flagellum	31 aa; TnlacZ/in	19/37	Smith et al., 2003; Barrett et al., in preparation
Flk/Salmonella IM	31 aa; TnlacZ/in or TnphoA/in	6/7	Aldridge et al., 2006
TraI/E. coli cyto	31 aa; TnlacZ/in	21/33	Haft et al., 2006
SpoIIIE/B. subtilis IM	31 aa; TnphoA/in	5/16	Lui et al., 2006

[a] Source refers to the organism where the mutagenized gene was originally isolated and the normal cellular location of the expressed protein.

[b] The definition of a "permissive" site for the purposes of this table is somewhat flexible. Here, a permissive site mutant is one that maintains at least 25% of the protein's primary activity, but some secondary activities may be compromised (e.g., the MalKi31 permissive site mutants are functional in maltose transport, but may not be proficient for various regulatory activities that the protein normally has). See the individual publications about the proteins for more details.

[c] The number of permissive/total insertional mutants isolated for these proteins is not clearly reported, but the number of characterized mutants is given.

Cyto, cytoplasm; IM, bacterial inner or cytoplasmic membrane; OM, outer membrane; PM, eukaryotic plasma membrane.

i31 mutants in LacI of *E. coli* (the *lac* repressor protein) (Nelson *et al.*, 1997). 12/18 LacIi31 mutants no longer fold into tetrameric complexes that can efficiently repress transcription from the *lac* promoter. Nevertheless, the majority of the mutants fall into expected classes from previous structure–function studies, and most of the mutants show that the i31 insertions yield phenotypes consistent with the location of the insertion. That is, the phenotypes of i31 mutants at particular locations are similar to amino acid substitution mutants at the same position.

Several studies have exploited stable but nonfunctional i31 mutants in different ways. Our laboratory has used i31 mutants of various Mal transport proteins to explore the functional interactions and the assembly pathway for this integral membrane protein complex. Studies with MalGi31 mutants in combination with binding protein-independent MalF mutants demonstrated a region in the third periplasmic domain of MalG that serves as a primary contact with the maltose-binding protein MalE during maltose transport (Nelson and Traxler, 1998). In contrast, particular nonfunctional MalFi31 and MalGi31 mutants (with their insertions into conserved cytoplasmic loops of those proteins) are deficient in MalK binding, but still associate efficiently with one another (forming a MalFi31-MalGi31 heterodimer) (Kennedy *et al.*, 2004). This analysis shows that these conserved MalF/MalG motifs are not important for the membrane proteins to associate with each other, and is consistent with the formation of a MalF-MalG intermediate as one route that can lead to the assembly of the final MalFGK$_2$ heterotetramer. More recently, Lui *et al.* (2006) have used functional and nonfunctional SpoEIIIi31 mutants to dissect different roles for the protein during *B. subtilis* sporulation. Their mutant data are consistent with the importance of this protein at multiple stages in this process, including chromosome segregation and translocation into the forespore and septal membrane fusion. Other examples can be found in the references provided in Table I, but the fundamental observation is that a collection of 10 to 20 i31 mutants can often be used to correlate a protein's interactions or activities to different domains.

Flexibility of Insertion Size and Sequence

For mutant collections derived from the Tn5-based transposon tools (such as Tn*lac*Z/in or TnTIN), the majority of the codons in the inserted sequence specify a unique amino acid sequence (for Tn*lac*Z/in and Tn*pho*A/in, 27/31 residues are the same in all mutants). In a few instances, we have investigated the influence of the particular sequence tag on the phenotype of the mutant protein. The tested parameters have included variations in insertion sequence and size (Gachelet *et al.*, in preparation). For instance, several transport-defective MalFi31 mutants were converted to smaller insertions, ranging from 2 to 21 residues (using Bal31 digestion, starting from the unique BamHI site in the insertional scar of the mutated gene). We isolated numerous smaller insertions that had widely varying amino acid sequences starting from three different mutants, but never found a Mal⁻ MalF insertion derivative that regained maltose transport activity.

We also asked whether we could isolate larger insertions in MalF at the site of permissive i31 tags, by mutagenizing *malF in vitro* with ISΩ/hah (Gallagher *et al.*, 2006), resulting in 51 codon insertions into the gene. This effort was less informative: we isolated several in-frame *malF*::i51 mutations (Table I) (Gachelet *et al.*, in preparation), including two in positions quite close to i31 insertions that allow robust expression of stable and active MalFi31 permissive-site mutants. However, none of the MalFi51 mutants were stable proteins in the cell. Despite this failure, work in our lab and elsewhere clearly demonstrates that some permissive sites are capable of tolerating substantially larger insertions of various polypeptide sequences beyond the 24 to 31 residue motifs, as described in the next section. In contrast to the generally good expression and stability observed for a variety of "smaller" i31 (or i24) mutants in both soluble and membrane-bound proteins, the stability/expression problems of MalFi51 mutants suggest that successful isolation of "larger" insertions of polypeptide sequences at permissive sites depends on (currently) unpredictable factors.

Engineering Permissive Sites with Other Functional Motifs

Several studies have demonstrated that at least some permissive sites identified in proteins (using both transposon tools and more traditional genetic and biochemical methods) will tolerate a wide variety of different insertion sizes and sequences. Previously published work has demonstrated that the *E. coli* outer membrane protein LamB has multiple permissive sites, at least one of which can be used to display an impressive array of polypeptide sequences up to 60 residues in length on the external surface of the cell without compromising the protein's proper localization and function (Charbit *et al.*, 1988). Brown (1997) exploited a LamB permissive site to identify tandem-repeating sequences of 14 residues (total insertion size of about 84 to 126 residues) that specifically bind to gold.

With these observations as a starting point, we have probed the flexibility of insertion size and sequence in permissive-site i31 mutants of the cytoplasmic TraI protein (the relaxase/helicase required for F-plasmid conjugation). The wild-type protein is large (1756 residues), and it has shown a striking capacity for additional polypeptide sequences at two different permissive sites. We have isolated stable and functional insertions of 53, 70, and 98 residues at permissive sites in the middle and near the C-terminus of the protein (Dai *et al.*, 2005; Gachelet and Traxler, unpublished observation). Notably, a 53-residue insertion near the C-terminus of TraI interacts at high affinity with Cu_2O (Dai *et al.*, 2005). The TraI-Cu_2O–binding derivative can interact simultaneously with Cu_2O nanoparticles and DNA *in vitro*. Other TraI permissive-site derivatives with the 70- and 98-residue insertions contain five and seven repeats of a 14-residue gold binding motif identified by Brown (1997). These proteins will also bind to gold nanoparticles and DNA *in vitro* (Przybyla *et al.*, in preparation).

Creation of Derivatives from Permissive i31 Mutants

A general procedure for inserting an additional sequence motif into the i31 sequence (as used for making the TraI-gold binding derivatives) takes advantage of a unique BamHI site located near the center of the i31 sequence and the compatible cohesive ends generated by BamHI and BglII endonucleases. We have used binding motifs (gold, Cu_2O, etc.) originally identified from combinatorial libraries and displayed within the context of some other protein. The sequences coding for these motifs can easily be PCR-amplified with primers that hybridize to the gene coding for the displaying substrate protein (LamB, M13 Gene III, or FliC, typically). A general strategy follows:

1. PCR-amplified sequence carrying BglII sites at each end is digested with BglII, while the plasmid with the permissive i31 mutation is digested with BamHI.
2. Ligation of the PCR fragment into the i31 site is performed using a large excess of PCR product (up to 18:1 insert:vector ratio), followed by a BamHI digest to restrict any plasmid that religated without the PCR-generated insert.
3. Ligation reactions are transformed into an *endA⁻* strain such as JM109, and plated on Luria agar supplemented with the appropriate antibiotic selective for the plasmid containing the i31 allele. Candidates are screened by PCR amplification, using primers specific for the new insertion. Roughly 20 to 30% of candidates typically carried the desired insertion.

With this strategy, the newly inserted material is incorporated within the existing i31 insertion. The flanking i31 sequence in our derivatives has not been a problem in our experiments. In fact, the i31 sequence may provide a desirable linker region between the new sequence motif and the displaying protein, so that the attributes of the additional insertion motifs do not interfere with the display substrate protein.

This methodology enables further engineering of permissive site mutants to increase their experimental potential. This can also be used for stable but nonfunctional sequence-tagged mutants with desirable characteristics (such as dominant-negative phenotypes). The sequence flexibility for additional sequence tags is high. The size range of permissible insertions is unclear, but likely falls between 20 and 100 (or more) residues in a context-dependent manner.

Acknowledgments

The work in the Traxler lab is supported by grants from the National Science Foundation (MCB#0345018 and DMR#0520567). We thank current and former members of the lab for helpful discussions and data, including Rembrandt Haft, Laralynne Przybyla, John Hodges, Angelica Talic, and Kristoffer Baek. We also thank Brad Cookson and Kelly Hughes for helpful suggestions and encouragement. We are especially thankful for continuing discussions and thoughtful comments from Colin Manoil and the members of his laboratory. Our work on engineering permissive site mutants for interaction with inorganic compounds has benefited terrifically from suggestions, advice, and scientific expertise from Stanley Brown (University of Copenhagen) and several members of the University of Washington Genetically Engineered Materials Science and Engineering Center.

References

Aldridge, P., Karlinsey, J. E., Becker, E., Chevance, F. F. V., and Hughes, K. T. (2006). Flk prevents premature secretion of the anti-σ factor FlgM into the periplasm. *Mol. Micro.* **60,** 630–642.

Brown, S. (1997). Metal recognition by repeating polypeptides. *Nat. Biotech.* **15,** 269–272.

Charbit, A., Molla, A., Saurin, W., and Hofnung, M. (1988). Versatility of a vector for expressing foreign polypeptides at the surface of gram-negative bacteria. *Gene* **70,** 181–189.

Dai, H., Choe, W. S., Thai, C. K., Sarikaya, M., Traxler, B. A., Baneyx, F., and Schwartz, D. T. (2005). Nonequilibrium synthesis and assembly of hybrid inorganic-protein nanostructures using an engineered DNA binding protein. *J. Amer. Chem. Soc.* **127,** 15637–15643.

Ehrmann, M., Bolek, P., Mondigler, M., Boyd, D., and Lange, R. (1997). TnTIN and TnTAP: Mini-transposons for site-specific proteolysis *in vivo. Proc. Natl. Acad. Sci. USA* **94,** 13111–13115.

Gallagher, L., Turner, C., Ramage, E., and Manoil, C. (2006). Creating recombination-activated genes and sequence-defined mutant liberties using transposons. *Methods Enzymol.* **421,** 126–140.

Haft, R. J. F., Palacios, G., Nguyen, T., Mally, M., Gachelet, E. G., Zechner, E. L., and Traxler, B. (2006). General mutagenesis of F plasmid TraI reveals its role in conjugative regulation. *J. Bacteriol.* **188,** 6346–6353.

Kennedy, K. A., and Traxler, B. (1999). MalK forms a dimer independent of its assembly into the MalFGK2 ATP-binding cassette transporter of *E. coli. J. Biol. Chem.* **274,** 6259–6264.

Kennedy, K. A., Gachelet, E. G., and Traxler, B. (2004). Evidence for multiple pathways in the assembly of the *E. coli* maltose transport complex. *J. Biol. Chem.* **279,** 33290–33297.

Lee, M. H., Kosuk, N., Bailey, J., Traxler, B., and Manoil, C. (1999). Analysis of F factor TraD membrane topology using gene fusions and trypsin-sensitive insertions. *J. Bacteriol.* **181,** 6108–6113.

Lippincott, J., and Traxler, B. (1997). MalFGK complex assembly and transport and regulatory characteristics of MalK insertion mutants. *J. Bacteriol.* **179**, 1337–1343.

Lui, N. J. L., Dutton, R. J., and Pogliano, K. (2006). Evidence that SpoIIIE DNA translocase participates in membrane fusion during cytokinesis and engulfment. *Mol. Microbiol.* **59**, 1097–1113.

Manoil, C. (2000). Tagging exported proteins using *E. coli* alkaline phosphotase gene fusions. *Methods Enzymol.* **326**, 35–47.

Manoil, C., and Bailey, J. (1997). A simple screen for permissive sites in proteins: Analysis of *Escherichia coli lac* permease. *J. Mol. Biol.* **267**, 250–263.

Manoil, C., and Traxler, B. (2000). Use of in-frame insertion mutations for the analysis of protein structure and function. *Methods* **20**, 55–61.

McMahon, C. W., Traxler, B., Grigg, M. E., and Pullen, A. M. (1998). Transposon-mediated random insertions and site-directed mutagenesis prevent the trafficking of a mouse mammary tumor virus superantigen. *Virology* **243**, 354–365.

Nelson, B., and Traxler, B. (1998). Exploring the role of integral membrane proteins in ATP-binding cassette transporters: Analysis of a collection of MalG insertion mutants. *J. Bacteriol.* **180**, 2507–2514.

Nelson, B. D., Manoil, C., and Traxler, B. (1997). Insertion mutagenesis of the lac repressor and its implications for structure–function analysis. *J. Bacteriol.* **179**, 3721–3728.

Smith, K. D., Andersen-Nissen, E., Hayashi, F., Strobe, K., Bergman, M. A., Barrett, S. L. R., Cookson, B. T., and Aderem, A. (2003). Toll-like receptor 5 recognizes a conserved site on flagellin required for protofilament formation and bacterial motility. *Nat. Immunol.* **4**, 1247–1253.

Waters, C. M., and Dunny, G. M. (2001). Analysis of functional domains of the *E. faecalis* pheromone-induced surface protein aggregation substance. *J. Bacteriol.* **183**, 5659–5667.

[10] Using Genomic Microarrays to Study Insertional/Transposon Mutant Libraries

By DAVID N. BALDWIN and NINA R. SALAMA

Abstract

The rapid expanse of microbial genome databases provides incentive and opportunity to study organismal behavior at the whole-genome level. While many newly sequenced genes are assigned names based on homology to previously characterized genes, many putative open reading frames remain to be annotated. The use of microarrays enables functional characterization of the entire genome with respect to genes important for different growth conditions including nutrient deprivation, stress responses, and virulence. The methods described here combine advancements in the identification of genomic sequences flanking insertional mutants with microarray methodology. The combination of these methods facilitates tracking large numbers of mutants for phenotypic studies. This improves both the

METHODS IN ENZYMOLOGY, VOL. 421   0076-6879/07 $35.00
DOI: 10.1016/S0076-6879(06)21010-3

efficiency of genome-saturating library screens and contributes to the functional annotation of unknown genes.

Introduction

Libraries of insertional mutants have long been useful tools for isolating genes important for specific functions in many model organisms, including *Drosophila melanogaster* (Cooley *et al.*, 1988), *Saccharomyces cerevisiae* (Kumar *et al.*, 2002), and many strains of bacteria (Judson and Mekalanos, 2000). Transposons have been widely used to generate such libraries for several reasons. Active transposons can be used for *in vivo* mutagenesis, generating large numbers of insertions fairly easily, and can be modified to be defective after a single round of replication, making single insertions per genome attainable. These transposons can be differentially tagged to facilitate tracking of different mutants in a pool, such as with signature-tagged mutagenesis (STM) (Hensel *et al.*, 1995). In addition, they provide a known target sequence to aid in identification of the insertion site.

Methods such as STM have contributed enormously to the study of pathogenesis by making possible the screening of large numbers of bacterial mutants *in vivo* (Shea *et al.*, 1996). One obvious advantage of screening large numbers of mutants simultaneously is the drastic reduction in the number of animals used. However, methods such as STM are still quite laborious due to the requirements for both amplifying and detecting regions surrounding individual insertions during screening. The approach described in this chapter combines much of the theory behind STM, with modern functional genomics tools and advancements in PCR-based identification of sequences flanking known insertions.

The primary methodology detailed here depends on the accurate description of mutants in a given starting pool, followed by characterization of the remaining mutants after treatment of the pool to some condition. This approach is a classic negative selection screen, where all the mutants are present at the beginning of an experiment. During exposure to a selective condition, mutants unable to survive the condition are lost from the pool, and do not reemerge in analysis of the output from the experimental condition. Genomic microarrays are used to define the mutants in both input and output pools simultaneously. Here we describe how to use a library of mutants and a microarray-based approach to search for all minimally essential genes in a microbial genome, as well as conditionally essential genes such as those critical for colonization and persistence in an animal model of infection.

The enormous advantage of using microarrays to track the behavior of insertional mutants is inherent in the randomness of the approach. One does not need to know anything about the mutants prior to the screen, and they can be tracked by their insertion location in large pools. Various methods have been developed for attempting this approach, pioneered with a transposon (Tn) containing a T7 promoter for transcribing regions flanking the Tn *in vitro* and subsequent labeling and hybridization to a microarray (transposon site hybridization, TraSH) (Badarinarayana *et al.*, 2001; Chan *et al.*, 2005; Sassetti *et al.*, 2001; Chapter 11 in this volume). More recently, approaches using methods of direct amplification for Tn-flanking regions with a PCR-based approach have proven very successful (Salama *et al.*, 2004), and we will focus on these as they are broadly applicable to existing transposon-based systems for mutagenesis. In fact, one could theoretically reanalyze old experiments using these modern tools.

An important advancement that made this approach possible was the advent of polymerase chain reaction (PCR) methods to amplify genomic regions flanking insertional mutants. This method was first described as thermal interlaced PCR (TAIL) for identifying T-DNA inserts in *Arabidopsis thaliana* (Liu and Whittier, 1995; Liu *et al.*, 1993, 1995), but was adapted independently as semirandom two-step PCR (ST-PCR) for the purpose of identifying transposon insertions in *S. cerevisiae* (Chun *et al.*, 1997). The basic idea is to use anchored random primers in combination with a primer specific to one side of the known insert in a first-round PCR reaction, followed by a second nested reaction for reamplification of Tn-flanking sequences. The PCR bands from the second reaction can be purified and sequenced in the case of individual mutant analysis. However, the power of this system comes from the ability to perform the reaction on collections of mutants, and to use the array for identifying all of them.

The combination of microarray methods with identification of Tn-flanking sequences by PCR is termed "microarray tracking of transposons" (Salama *et al.*, 2004). This chapter seeks to provide some general background for this approach, but more thoroughly to highlight the technical details of each step. There are many steps where the experimenter should consider optimizing conditions for each specific system, due to differences in the format of the microarrays and differences in the G/C content of the organism. Additionally, we include some important tips on how to increase the general effectiveness of microarray analysis. Lastly, we discuss the data acquisition, storage, and analysis of such genome-wide experiments in the context of web-based databases. Most researchers will find that this is a necessary component of microarray work, and such databases are useful not only for initial storage and analysis, but for later making the work publicly available.

With many bacterial genes still unclassified, these approaches will greatly increase the pace of functional annotation. Public availability of functional genomics experiments will allow cross-referencing or "meta-analysis" to assign or eliminate potential gene function.

Step 1. Generating Insertional Mutant Libraries

The first step of course is to generate a library of mutants. The methods for this are highly dependent on the specific organism, since the genetic tools for introducing transposons vary substantially. Here we focus on the use of a mini-Tn7 cassette in the bacterium, *Helicobacter pylori*. While for other bacteria "live" transposons can be transduced by infecting with carrier phage, or introduced on a plasmid by transformation or conjugation, these tools do not exist as such for *H. pylori*. Instead, insertions were generated *in vitro* using purified chromosomal DNA, a plasmid DNA bearing the miniature transposon Tn7 with a chloramphenicol-resistance cassette, and recombinant transposase (New England Biolabs) (Salama *et al.*, 2004). Following the *in vitro* reaction, the recombinant (mini-Tn7–containing) chromosomal DNA was introduced by natural transformation back into the original strain of *H. pylori*, and individual mutants were selected on chloramphenicol plates. For any library in any organism, the final step is very important, namely to ascertain the number of independent mutants (i.e., the complexity of the library). This is done simply by counting transformants (transposition events in some cases) in the original plating before amplification of the library for use in a screen. It is also of interest to check individual clones from any library to ensure that they contain only a single insertion, as well as to ascertain that "hot spots" are not dominating either the recombination or the transposition events. While it is unrealistic to check all the clones in the library, it is feasible to check a significant number (20 to 40) by Southern blot analysis using a unique site in the Tn cassette. When a library of sufficient complexity is generated, it can be harvested directly by scraping sufficient numbers of single colonies from a large number of plates, mixed and stored in aliquots at –80°. Single colonies are often preferred for library amplification since any growth advantages/disadvantages of individual mutants cannot overwhelm the final culture.

Step 2. Screening Mutant Libraries: A Primer

Again, this aspect of the protocol will vary substantially based on the organism in question. For *H. pylori*, we detail two examples for how the microarray has been useful in screening the genome with transposon mutants. First, we analyzed a large number of mutants to define genes whose mutagenesis could be tolerated without growth defect. The corollary reveals the essential genes of the organism, namely those that could not be mutagenized. For a genome of about 1600 genes, a library of 10,000 individual clones was considered nearly saturating. If the library is indeed saturating, implying that every gene has been mutagenized multiple times, then genes that never appear mutagenized during analysis are presumably essential to the organism under plating conditions (see section titled "Data Analysis: Minimally Essential Genes" below).

The second application describes analysis of conditionally essential genes (Fig. 1). These genes do not play a major role in housekeeping or biochemical maintenance of the organism, but are only required during certain aspects of the bacterial lifestyle. We focus on methods for identifying conditionally essential virulence genes, that is, genes required for establishing an infection *in vivo*, but not essential for growth on plates (see "Data Analysis: Minimally Essential Genes" section).

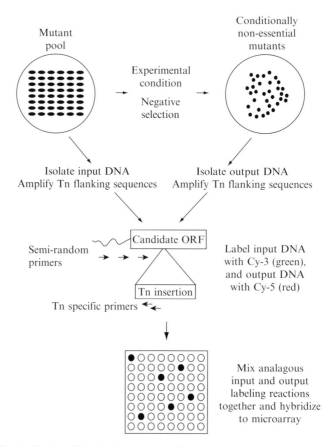

FIG. 1. Basic outline of microarray tracking of transposon mutants *in vivo*.

Step 3. Microarray Analysis of Mutant Pools and Mapping
   Transposon Insertions

Basic Experimental Flow

Bacteria are plated for single colonies or patched either as starting pools, or output pools from an experimental condition. (Harvesting DNA from colonies or patches helps normalize the amount of starting DNA for each clone in the pool.) Genomic DNA is prepared from the pool of mutants, and diluted for Tn insertion analysis. Starting pool mutants can be identified either with single-color fluorophor labeling (Cy-3, emits at 532nm, "green"; Cy-5, emits at 635nm, "red"), or in comparison with the output from a given experiment using a two-color scheme. In either case, two-step nested PCR reactions are performed to specifically amplify regions surrounding the transposons and to label the DNA neighboring the transposons for micro-array analysis (Fig. 2). By labeling each side of the Tn in a single-color reaction, one can either compare the two sides directly to help define the exact location of an insertion (red/green=Tnleft/Tnright), or compare the same side of the insertions in a pool of mutants before and after an experimental condition (red/green=Tnleft$_{after}$/Tnleft$_{before}$).

In the first PCR reaction, anchored random primers (e.g., 5′-nested/fixed sequence-N$_{10}$-GATAC-3′) are used in combination with a Tn-specific primer to semispecifically amplify flanking sequences. The second nested reaction uses one primer designed to anneal to a fixed 5′ portion of the randomly anchored primers, and a second Tn-specific primer internal to

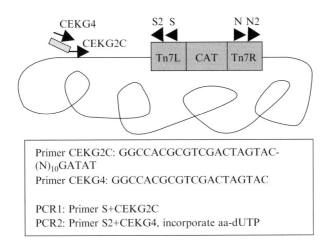

Primer CEKG2C: GGCCACGCGTCGACTAGTAC-
(N)$_{10}$GATAT
Primer CEKG4: GGCCACGCGTCGACTAGTAC

PCR1: Primer S+CEKG2C
PCR2: Primer S2+CEKG4, incorporate aa-dUTP

Fig. 2. Semirandom PCR for amplifying and labeling DNA adjacent to transposons. Reproduced from Salama *et al.* (2004) with permission form the American Society of Microbiology Journals Department.

the primer used in the first reaction (Fig. 2). Labeling the Tn-containing DNA with fluorescence dyes is achieved by incorporating amino-allyl dUTP as a fraction of the dTTP in the second PCR reaction, and then chemically coupling monoreactive fluorescent dyes in a final reaction. The DNA is then either hybridized directly to the microarray (in the case of single-labeling experiments), or mixed with the labeled output DNA and hybridized (in the case where experimental samples are being compared to starting pool DNA, or a reference DNA pool).

Finally, microarrays are washed, dried, and scanned to collect the data for individual spots representing genes that either contain or are adjacent to transposon insertions. The data can be stored and analyzed in a number of different ways, and for the sake of example we will describe the storage of a large data set in a web-based relational database (Stanford Microarray Database, SMD; see Sherlock *et al.*, 2001). The obvious advantage of a system like SMD for storage is that the data can be retrieved from any Internet-capable computer. However, as we will see, the individual experiments are typically analyzed as such in individual spreadsheets, and then collated in the case of a genomic screen. Analysis will depend on the number of mutants in pools being analyzed, the number of spots per gene on the array, and so on.

Microarray Design

Because many types of microarray now exist, we will focus on the use of PCR-based spotted arrays. Much consideration should be taken in designing primers to amplify open reading frames (ORFs) or intergenic regions for printing a microarray. The size of the products should be limited to the range of 300 to 1000 bp, and primers should be grouped according to the annealing temperatures for high throughput amplification. In addition, one may want to consider including a number of control spots that do not represent ORFs of the organism, such as genomic DNA or plasmid DNA containing known antibiotic-resistance genes that might be useful in later analyses. Having heterologous genes such as antibiotic-resistance genes on the array helps with optimization of both single gene identification as well as gene expression studies where *in vitro* transcribed RNA can be doped into samples as controls. Genomic DNA spots can be useful for normalization of the two-color channels during data collection (scanning) and later analysis. One might also want to consider printing each ORF or intergenic region multiple times on each array (if space and printing material are not limiting) for the purpose of examining reproducibility within an experiment.

Preparation of Bacterial Genomic DNA

First, it should be noted that not all bacteria are created equal with respect to protecting their genomes from the lytic procedures of DNA isolation. Some optimization will be required depending on the cell wall components, but in general the bacteria are dissolved in a strong denaturant, followed by precipitation of proteins, and finally precipitation of nucleic acids from the resulting supernatant. The trick is quite the opposite of plasmid preparation, where chromosomal DNA is deliberately left coupled to proteins and membranes during precipitation. In these methods, releasing the chromosomal DNA from its binding partners is essential for good yield.

CTAB Method

Pellet 1.5 mls of bacteria from broth, resuspend in 567 μl 10 mM Tris.Cl, 1 mM EDTA pH 8.0, add 3 μl of proteinase K (20mg/ml) and 30 μl of 10% sodium dodecyl sulfate (SDS) (0.5% final). Incubate at 37° for 1 hour. Add 100 μl of 5 M NaCl, vortex, and add 80 μl CTAB/NaCl (10% cetyltrimethylammonium bromide, 0.7 M NaCl) and incubate at 65° for 10 min. Add an equal volume of chloroform/isoamyl alcohol, vortex, and microcentrifuge for 10 min at highest rpm. Transfer the supernatant to a fresh tube, and extract with an equal volume of phenol/chloroform/isoamyl alcohol to remove all CTAB and microcentrifuge at high speed. Transfer supernatant to a fresh tube, add 0.6 volumes isopropanol, and invert gently to precipitate genomic DNA. Microcentrifuge at high rpm, and wash DNA pellet with 70% ethanol. Remove all remaining wash solution, and resuspend DNA in 1–200 μl TE for quantitation (Ausubel *et al.*, 1997).

Alternative Method

This protocol is available as a part of the Wizard Genomic DNA Purification Kit from Promega. It is similar to the traditional alkaline lysis procedure for isolation of plasmid DNA and is adaptable for Gram-positive bacteria and Gram-negative bacteria. Pellet 1.5 ml of bacteria, and resuspend in 600 μl nuclei lysis buffer (detergent). Vortex until bacteria are completely resuspended, and then incubate at 80° for 10 min. Add 3 μl of RNAse (10 mg/ml in TE) and incubate at 37° for 30 min. Add 200 μl protein precipitation buffer (high salt/acetate buffer) and place on ice for 30 min. Microcentrifuge at high speed for 10 min, transfer supernatant to a fresh tube, and add 600 μl isopropanol. Precipitate nucleic acids by inverting several times, followed by microcentrifugation at high speed. Wash DNA pellet with 70% ethanol and spin again, being sure to remove all excess liquid. Resuspend pellet in 200-μl TE for quantitation.

Amplification of Transposon-Containing DNA: First-Round PCR

The goal of the first round of PCR is to amplify sequences flanking one side of the transposon in each mutant from a given pool. Flanking sequences are amplified using a semispecific approach where half the primer pool is a single oligonucleotide complementary to one arm of the transposon, and the other half of the primer pool is a mixture of random primers (length of random portion is up to the individual researcher based on empirical results, but the range should be 6 to 10 random nucleotides) with a conserved five-nucleotide anchor at the 3' end (anneals roughly once every kilobase depending on G/C content of the anchor and genome), and a second conserved sequence at the 5' end to serve as a template for the second-round PCR (nested reaction; see Fig. 2) (Manoil, 2000).

In the first round, 20-μl PCR reactions are performed as follows:

2 μl 10× buffer (specific to the Taq polymerase used)
2 μl dNTPs (2.5 mM)
1 μl Tn specific primer (40 μM)
1 μl of anchored random primer (40 μM)
2 μl of genomic DNA (50 ng/μl)
0.5 μl Taq
11.5 μl dH$_2$O

Thermocycling is programmed such that the template is initially denatured for 2 min, and then six cycles are performed where the initial annealing temperature is 42° for 30 sec; with each successive cycle, the temperature drops 1°. Extension is performed at the usual 72° for 3 min. These first six cycles permit the anchor and random sequence to anneal efficiently. Subsequently, 25 cycles are performed where the annealing temperature is fixed at 65°, encouraging further amplification of flanking sequences. Sequences amplified in this reaction almost never appear as a single band on a gel. If the reaction is efficient, one might see a faint smear, but so little product is typically amplified from any one Tn that almost nothing is visible. These fragments do, however, serve as an excellent template for the second, nested reaction where specific fragments can be amplified and labeled.

Reamplification and Incorporation of aa-dUTP into Transposon-Flanking DNA: Second-Round PCR

The goal of the second round of PCR is twofold. First, it is important to reamplify the Tn-flanking sequences in a nested reaction, ensuring that the majority of fragments labeled for microarray hybridization contain Tn sequences, and not random primer sequences at both ends. Second, this reaction facilitates incorporation of amino-allyl dUTP, enabling fluorescent labeling of products being reamplified. Some experimenters may choose to incorporate fluorescently labeled nucleotides directly in this reaction, in which case they can skip directly to the step of purifying the PCR product for hybridization. In addition, one might consider using fluorescently labeled primers specific to the transposon, further improving specificity for flanking sequences. However, one might encounter sensitivity difficulties with this approach since each PCR product could contain a maximum of labeled nucleotides contained in the primer, in contrast with labeling throughout the product. Another approach to improving specificity is to limit the length of the labeled product by restriction enzyme digestion of either the starting genomic DNA, or the resulting cDNA. Products less than 1 kb can be size selected for labeling, improving the sensitivity of hybridization to flanking regions, as well as permitting much larger pool size analysis due to less overlap with neighboring ORFs. This method is described in detail for the use of transposons containing transcriptional fusions, such as those incorporating the T7 RNA polymerase promoter used in TraSH (Sassetti and Rubin, 2002). It should be noted that in this case, genomic DNA is fractionated prior to any amplification or labeling.

The second-round PCR is set up as follows: The first-round product is first diluted with 80 μl of dH_2O to a final volume of 100 μl, in effect diluting out the original template and primers. The second-round PCR reaction is designed to generate ample amounts of product for hybridization, in a final volume of 100 μl.

1 to 2 μl diluted first-round template
10 μl 10× PCR buffer
10 μl 3-mM dNTP/aa-UTP (3 mM dA,G,CTP, 1.2 mM dTTP, 1.8 mM amino-allyl-dUTP (aa-dUTP))
1 μl nested primer to the fixed end of the first-round PCR product (40 μM)
1 μl nested Tn specific primer (40 μM)
1 μl Taq
76 μl dH_2O

The theromocycler is programmed for a more traditional reaction in the second round, where after denaturing the template for 2 min at 94°, 30 cycles are performed by annealing at 56° for 30 sec, extending at 72° for 2 min, and finally denaturing again at 94° for 30 sec.

Purification and Labeling of Second-Round PCR Products

A number of methods are available for purifying, concentrating, and labeling the second-round PCR products for hybridization to microarrays (the labeling step can be omitted if precoupled fluorescent nucleotides are used in the reaction). In the case of labeling the PCR products with monoreactive dyes (Cy-3, green; Cy-5, red) after the second-round reaction, two steps are critical. The first is to clean up the DNA by removing free amino groups (and nucleotides) while concentrating the cDNA, and the second is to efficiently mix that DNA with the monoreactive dyes in the appropriate buffer for coupling to the amino-allyl group of the aa-dUTP. In this protocol, we describe the use of Zymo Research DNA Clean and Concentrator columns, since they do not release any particulate matter into the DNA prior to hybridization, and recovery of the PCR product is excellent.

PCR products are mixed with 300 μl of DNA-binding buffer (provided in kit, and according to manufacturer's instructions; a minimum of two volumes) and loaded on the spin column. After spinning and discarding the flow through, the column is washed two times with a salt/ethanol wash buffer and spun dry. The column is then transferred to a microcentrifuge tube containing an aliquot of dried monoreactive dye. (Individual mono-reactive dyes can be purchased from Amersham Biosciences in aliquots sufficient for labeling 1 mg of protein. This is approximately 20 times what one needs for labeling an individual PCR reaction. To aliquot, resuspend the dye in 200 μl of dH$_2$0, aliquot into 20 individual prelabeled tubes, with 10 μl into each tube. Dry down in a speedvac, close tubes, and store at –20°). The second-round PCR products are then eluted with 15 μl of 50-mM sodium bicarbonate, pH 9.0, by spinning directly from the Zymo column onto the dried aliquot of dye. (With other systems for purifying the PCR product, you may need to elute with water, and mix the product 50:50 with 100 mM of sodium bicarbonate before mixing with the dried dye for the final labeling reaction). The dye is resuspended with the eluted DNA and allowed to couple in a dark drawer for at least 30 min (this reaction can proceed for as long as 8 to 12 hr). Some protocols suggest quenching the reaction with 4-M hydroxylamine when it is complete, but this is not necessary when using the Zymo purification.

Purification and Preparation of Labeled PCR Product for Hybridization

There are several important elements in this procedure. Eliminating all of the unincorporated dye from the labeling reaction is critical to avoid background problems during the hybridization, and it is also very important to add the hybridization constituents in the appropriate order to prevent precipitation of SDS, another source of background problems during scanning. The eluted DNA should never be refrigerated or frozen after addition of the SDS because some small fraction of the inevitable precipitate will never go back into solution in the presence of SSC and DNA, and the resulting arrays will be bright green with background.

Bacterial microarrays, which are small in comparison with mammalian microarrays, require fairly small hybridization volumes. The conditions described here are for an array fitting under a flat, square, 22-mm cover slip, with a final hybridization volume of 15 μl. Volumes should be adjusted accordingly if you are working with bigger arrays.

The labeled DNA is purified by repeating the Zymo purification (see previous section), and eluting the labeled product with 12 μl of 10-mM Tris, pH 8.0. For two-color experimental schemes, the appropriate Cy-3– and Cy-5–labeled samples can be mixed in the same DNA binding buffer and co-purified without quenching or deleterious effects. The labeled and purified PCR products are then mixed carefully and sequentially with 1 μl of yeast tRNA (25 mg/ml), 1.5 μl of 20× sodium chloride/sodium citrate (SSC), and finally 1.5μl of 1% SDS. Immediately prior to hybridization, samples are heated to 95° for 3 min to denature labeled products, microcentrifuged to spin down any condensation on the lid, and gently remixed.

Hybridization and Washing Spotted Microarrays

Carefully pipet the hybridization mixture into the center of the microarray, being sure not to leave any bubbles on the surface. Bubbles can be popped *in situ* with a needle if necessary prior to placement of the coverslip. Occasionally, if the coverslip is dropped too quickly, bubbles will form underneath it. Sometimes these can be removed by gently pressing the coverslip on one side, but it is impossible to remove the coverslip and start over. Often, the bubble will expand and disappear during the hybridization, so it is better to leave it and see what happens than to try and get rid of it. Before closing the hybridization chamber, place several small volumes of 3× SSC on the microscope slide, but distal to the array. This will prevent drying under the coverslip during the incubation. Place the chamber in a water bath set to the desired hybridization temperature. This will depend on the G/C content of the genome, the type and size of DNA printed on the array, and so on. Controls should be performed to examine the optimal temperature for this step. Timing is also a variable, and should be studied with the array being used. The standard is to leave the chamber at a specified temperature overnight, which typically means 10 to 12 hr minimum. This is aimed at bringing the hybridization reaction to equilibrium. Some researchers prefer to use special elevated coverslips that they claim can shorten the hybridization time by permitting better fluid dynamics. Fancy chambers and coverslips are now available that permit rotation of the mixture in an oven, instead of submerging in water baths. Engineering progress in this field is extremely rapid. Some also claim that if enough moisture is present in the chamber to prevent drying, hybridization can occur over a whole weekend. You decide, but do not cut it too short.

Washing the arrays is performed by submerging them in a succession of increasingly stringent washes of 5 to 10 min each, starting with

$2\times$ SSC/0.1% SDS (it is critical that the SDS be completely in solution), and then $2\times$ SSC with no SDS, $1\times$ SSC, and if necessary, $0.2\times$ SSC. If the array is placed carefully in a slide holder, such that the coverslip is on the underside as it tilts in the holder, the coverslip will gently fall away from the slide in the first wash. To be most efficient at washing away the remaining SDS (critical), it is best to transfer the individual arrays, now free from the coverslip, to a new holder in the second wash, carefully blotting the edge of the array on a Kimwipe as you transfer to remove all excess liquid. After the last wash, carefully balance a low-speed centrifuge (with adapters for microtiter plates) with another slide holder, placing blotting paper underneath the slide holders. Spin the slide holders at 500 to 800 rpm for 5 min to remove all wash buffer. It is important that the centrifuge be clean prior to this drying step. Any dust, ethidium bromide, media, or organic solvent present in the centrifuge will end up on the array during the spin, ruining a lot of hard work. After drying, slides should be stored in the dark whenever possible to avoid small amounts of fluorescence quenching.

Data Collection and Normalization

Fluorescence scanners for microarrays are now widely available, made by a number of companies such as Axon (now Molecular Devices), Perkin Elmer, Alpha Innotech, and Agilent, among others. Most come with their own software for both collecting the array image from the scanner, and for collecting the data for individual elements on the array by defining the spots and in turn the signal within and around them. We will not detail the use of any particular device or software here. However, there are a few important things to remember when acquiring data from microarrays in general. For the most part, these concepts apply to arrays being used with a two-color scheme, where one "condition" is being compared to another. In addition, the experiments we are describing in this chapter require additional consideration since one might only expect a fraction of the spots on the array to give any signal at all.

Most statistical methods for analyzing microarray data rely on the notion that both channels (Cy-3, Cy-5) are represented equally with respect to their intensity and distribution across the array in any given experiment. This is the case for comparing two genomes, or comparing the transcriptional state of an organism under two different growth conditions. When scanning, one should try to create intensity distributions in the data that reflect this. Such distributions can be achieved by watching the data collection in real time, and adjusting the photomultiplier tube (PMT) settings accordingly, until the data for the two channels are roughly overlapping with respect to intensity and distribution. This is not an exact science by any means, and the data can be further "normalized" (the distribution of data for each channel can be made to overlap with each other exactly by calculating a normalization factor for either the mean or median of one channel and applying it to the other channel). It is also important to get as much dynamic range in the data as possible. This is accomplished by setting the PMT settings as high as possible without saturating any spots (forcing them to the limit of the data range). In general, it is prudent to find the limit of saturation first, set the PMTs below that, and then to visually normalize the distributions prior to final scanning (one can watch the histograms of the data for each channel while scanning in real time).

The concept of normalizing the two channels to each other is widely accepted in the genomics field; however, it is important to consider the nature of the experiment in each case. For the experiments described here, where we are trying to locate 50 individual insertions in a pool, only a fraction of the spots (ORFs) on the array will give any signal. In a screen where a significant fraction of mutated genes have a phenotype, the resulting arrays may be slightly dominated by one channel or the other. One method for solving this problem is to include a large number of control spots in the design of the array, such as genomic DNA spots. Instead of normalizing the distribution of all spots on the array, one can normalize the channel intensities by the spots that should contain roughly the same amount of signal or some expected signal. In considering this before printing, you should also consider placing some of these spots in each block of the array (use each tip of the arrayer to print some of the control spots).

Data Storage

In any genomic screen, data storage can be difficult. It gets especially complicated with microarray data, because the output spreadsheet from any given array will contain thousands of rows of genes and many columns of information about the data for each spot on the array. The most useful tools for storing microarray data are relational databases. One excellent example of such is the Stanford Microarray Database (SMD) (Sherlock *et al.*, 2001). Each different category of data for each array is partitioned into the database, and can be accessed independently. One can simultaneously apply a wide range of filters to the data when retrieving it from the database. In the case of SMD, a web-based relational database, data can be retrieved from any computer with Internet capability. A major advantage is that one can repeatedly apply new criteria/filters to the data without ever fearing the loss or alteration of the original data. One can also choose the type of data to retrieve and filter for any spot, and it may be relevant to access both raw channel-intensity data as well as ratio data for both channels. For some statistical applications, it may be necessary to retrieve raw and normalized data by the same filtering criteria, and then store outputs from analysis in a personalized database, such as an Access database (Microsoft).

Data Analysis: Screening for Minimally Essential Genes

Data analysis for any microarray experiment will depend on the details of the experiment itself. First we consider the use of microarrays to study the initial library of mutants, and what information might be contained in the genes that could not be mutagenized. In these experiments, large pools of mutants are studied to describe genes that can be mutagenized without a resulting growth defect.

In the experiments described here, the microarray itself has two spots printed for each ORF, and a series of genomic DNA spots for normalization of input and output pool channel intensities. While the genomic DNA spots serve as a useful starting place for normalization during the scanning process, standard normalization by mean/median channel intensity from the distribution of the whole array works well as long as enough spots pass as "good spots" by the regression correlation analysis. Regression analysis of individual spots serves as a filter for spot quality by demanding that the pixels within each spot are similar to each other above a threshold correlation value (determined by the user, and dependent on both the quality of array and the material to be labeled).

In the case of identifying insertions across the entire library, large pools can be used since they are not subject to the same severe stochastic effects of an experimental condition such as those encountered during *in vivo* analysis. To characterize the distribution of insertions in the entire library by identifying the hits and "non-hits" (i.e., essential genes) in *H. pylori*, 20 pools of 300 mutants were used. The goal of the approach was to label both sides of the transposon in separate color reactions, and then combine them in order to see which genes appeared in both reactions (implying the transposon is near the center of a gene), or in only a single reaction (implying that the transposon lies near the end of a gene, or between genes).

For each transposon pool, amplification from the left side of the transposon using primers S and S2 was labeled with Cy-3 (green [G]) and amplification from the right side of the transposon using primers N and N2 was labeled with Cy-5 (red [R]). To identify chromosomal loci that contain transposon insertions in each pool, the data were analyzed as follows to assign each gene a probability of containing an insertion (P[insertion]). We first averaged the logarithm base 2 of the red/green ($\log_2 R/G$) signal for duplicate spots on the array for each gene. These values were arranged in chromosomal order. Genes containing a transposon insertion would generate a signal in both channels, producing a yellow spot defined as $-1 <$ $\log_2 R/G < 1$ and having a P[insertion] value of 1. To account for cases where the transposon was at the very end of a gene, we identified gene spots for adjacent genes in the chromosome, containing strong signal in opposite channels defined by the following formulae: $gene_i \log_2 R/G < 1$ and $gene_{i+1} \log_2 R/G < -1$ or $gene_i \log_2 R/G < -1$ and $gene_{i+1} \log_2 R/G > 1$, where i equals gene order number. In such cases, each gene was given a P[insertion] value of 0.5. To calculate the number of insertions per gene, the number of insertions was summed across all 20 pools and then rounded down to the nearest integer. Genes that never appeared to contain hits in this analysis were studied further to determine whether they were in fact essential (Salama *et al.*, 2004).

Data Analysis: Screening for Conditionally Essential Genes Such as Colonization Factors

The analysis of smaller pools, such as those used in the search for conditionally essential genes *in vivo*, is slightly more complicated. In our case, we have analyzed *H. pylori* infection of the stomach mucosa (Baldwin *et al.*, 2006). Consistent with other observations of bottlenecks to colonization during mucosal infections, we found that with large pool sizes, the independent action of clones lost some reproducibility. In these experiments, pools of 48 mutants were chosen after comparing various pools sizes *in vivo*. Mutant pools were created to contain equal concentrations of each mutant by picking and patching the original library from single colonies onto gridded agar plates. When patches were all grown to a similar density, plates of 48 mutants were harvested and stored at $-80°$. These pools were amplified slightly as single colonies during the thawing procedure, and then used to infect mice or prepare "input" genomic DNA. Mice were infected for either 1 week or 1 month, after which stomachs were homogenized and plated for amplification of the remaining bacteria. The goal in analyzing each individual experiment, or condition that the pool of 48 mutants has been subjected to, is to identify the mutants both in the starting pool, as well as the mutants still present in the pool after treatment. In this example, the green channel (Cy-3, 532 nm) labels and identifies the mutants from the starting pool, and the red channel (Cy-5, 635nm) to describe the mutants still present in the output pool. Therefore, the green channel alone can be used to select the genes that originally contain transposon insertions, and the ratio of the red/green, or the LogR, can be used to assess the fitness of those mutants in the context of the experimental condition.

Data both for mutants represented in the input pool and the *in vivo* fitness of each mutant within the pool is contained in each array, each of which is derived from a single experimental mouse infection. The first goal is to identify the mutants contained in each input pool. The brightest spots in the green channel represent genes either containing or neighboring transposon insertions in the input pool. Describing the brightest green channel spots can be done in several ways (e.g., by sorting or ranking spots). We chose to use the filter described below.

1. The data for each array were retrieved from the database using a basic filter for data quality (the regression correlation of pixels within individual spots must be greater than 0.6).

2. We sorted the data by the green channel intensity, and temporarily removed the brightest 100 spots while calculating a mean intensity for all the remaining dim spots (we calculated the mean-100 because each gene is printed twice, and our pools contain roughly 50 mutants).

3. For each array, we then selected green channel signals that lie four standard deviations from the mean of the dim spots. This calculation returned varying numbers of spots (mean number of spots was 150; range 84 to 164), depending on the success of the PCR and labeling of the input pool. Importantly, the data for the ratio of the output/input pools could be carried through the analysis, and provided the measure of fitness for each gene considered present in the input pool.

In our experiments, two data sets were collected for each experiment (mouse infection): one in which the amplification of flanking DNA was performed from the left side of the transposon (S primers), and a second where the amplification of flanking DNA was performed from the right arm of the transposon (N primers). We used information from both sides of the transposon to predict whether the insertion was in the middle of a gene, at one end of a gene, or in a neighboring gene.

To call an insertion site within a gene, we required that its gene spot be labeled from both sides of the transposon insertion, and thus data for the same gene spot were collected in both data sets. These genes were assigned a probability of containing an insertion of 1.0. To account for transposon insertions where the insertion lay near the end of a gene, we also called insertions where after arranging the gene spots in chromosomal order, there was signal for a gene spot in the reaction from one side of the transposon and signal for the nearest neighboring gene spot in the reaction from the other side of the transposon. In this case we gave each of the two neighboring genes a probability of insertion of 0.5.

We then used the LogR data for each gene to assign either positive or negative values to the above probabilities to facilitate ranking genes with the most reproducible phenotypes. Genes where insertion had no effect on colonization, that is, if the spot was yellow (present in input and output, $\log_2(R/G) > -2$), we assigned a positive value to the probability. If the

LogR data indicated that the spot for a gene was green (present in input, absent in output, $\log_2(R/G) < -2$), we assigned the probability score a negative value. Insertions with no apparent phenotype (present in both input and output) received a value of 1.0 or 0.5 and insertions with a mutant phenotype (absent in output) were given a value of –1.0 or –0.5. The data for all the pools tested in each strain background were analyzed separately. For each strain background, the data were summed across all the pools and genes with a value less than –1.5 (a value of –0.5 in each of three mice) were considered candidate mutants.

One common source of stochastic error, or variation, occurs during the labeling, data collection, and normalization of the arrays. For example, in the experiments looking for conditionally essential genes for *H. pylori* during stomach colonization, we did not expect to find genes whose mutagenesis would improve fitness of the bacteria in the stomach. From one day to the next, however, we regularly saw a few spots on the arrays that appeared red during scanning. This could have been due to the fact that we were always using the same total amount of DNA for each labeling reaction, yet we had lost as many as half the clones in the pool during selection in the stomach. In addition, the mutant clones capable of colonizing the stomach almost certainly did not colonize at identical levels, contributing further to variable distribution in the output (plating bacteria from the infected stomachs). In the end, true stochastic effects were eliminated during the analysis by the stringent criteria of insert identification over a series of experiments. In the final analysis, no gene ever appeared to improve fitness of the organism when identified from both sides of the transposon, despite appearing red every time under one condition, such as a specific primer pair.

In contrast, one caveat to our chosen stringent criteria was that we undoubtedly overlooked some real insertions during the mouse infections. This was partly due to arbitrary cutoff limits during analysis, partly due to insert location in the genome, and also partly due to the primers we chose to amplify regions flanking the insertions. These criteria were chosen so that we could be highly confident of what we chose to pursue or report, but we acknowledge that within the data set, there is probably quite a bit more information about clones in each input pool.

Discussion and Conclusions

One major advantage to using a microarray-based approach in screening mutant libraries is not only the rapid and accurate identification of mutant loci, but also the high-throughput nature of the experimental design. Many mutants can be tested and identified simultaneously with a single array experiment, depending on the stochastic factors related to experimental design. Pool sizes should always be tested empirically to optimize for individual experimental conditions. When stochastic effects appear regularly, one should either reduce the pool size, or one should repeat the biological experiment enough times to generate a statistically relevant data set for each pool. Either way, the advantage in this approach manifests in the requirement for many fewer experiments than the number of clones in each pool.

The screen for conditionally essential genes described here is susceptible to the same caveats of many other genetic screens. The array is useful for localizing Tn insertions, but even predicted phenotypes from experiments described here can be misleading. It is possible to have insertions in intergenic regions that both do and do not affect expression of a neighboring gene product. It is also possible to have insertions within genes that do not affect function of the gene, even if the gene is essential under the conditions tested. Perhaps an insertion in an essential gene does not actually affect the function of that gene due to its location, but instead has severe polar effects on a neighboring gene. These are all problems one should be aware of while picking candidates for further analysis. Another source of error might be the result of cross-complementation of mutants in a given pool, or across a set of pools during analysis. Perhaps the wild-type bacteria (or nonessential mutant) makes a secreted factor that is protective of its own mutation in other clones in the stomach. Our method of prioritizing mutants from the entire screen could lead to cancellation of a legitimate mutant phenotype from one pool by a null phenotype from an intergenic or neighboring gene insertion. In general, this approach is very successful at predicting insertion sites, as well as strong phenotypes, but we acknowledge that there are numerous caveats to using these data to say that a gene has been mutagenized, but is not essential under a specific condition.

Of course it is critical to follow up these initial microarray results and to verify not only that the insertions are where one thinks they are, but also to verify that the predicted phenotype is attributable to the predicted insertion. In our case, inserts were verified by sequencing, and isogenic mutants were made with knockout cassettes for individual genes by straightforward PCR methods. Mutants derived from homologous recombination events were retested both alone (to determine the ID50, the dose of bacteria at which 50% of mice are infected) and in direct competition with wild-type bacteria (to determine the competitive index of the mutant, the number of bacteria that survive in direct competition with wild-type bacteria). The type of experiment one might use to verify an array result will depend entirely on the organism and system of choice.

The methods described here are very powerful for analyzing large collections of insertional mutants. While we have focused on our system with *H. pylori*, this approach is broadly applicable to any system where collections of insertional mutants might be of interest. One could study target sites of gene therapy vectors in collections of human cells, mariner insertions in *Drosophila*, or transposons in *Arabidopsis*. As microarrays for all sequenced organisms become available, these methods or adaptations thereof will prove more and more useful for rapid identification of genes involved in specific biological processes.

Acknowledgment

Work in our library was supported by Public Health Service Grant AI054423 from the National Institute of Allergy and Infectious Diseases.

References

Ausubel, F., Brent, R., Kingston, R. E., Moore, D. D., Seidman, J. G., Smith, J. A., and Struhl, K. (Eds.) (1997). "Short Protocols in Molecular Biology." John Wiley & Sons, New York.

Badarinarayana, V., Estep, P. W., 3rd, Shendure, J., Edwards, J., Tavazoie, S., Lam, F., and Church, G. M. (2001). Selection analyses of insertional mutants using subgenic-resolution arrays. *Nat. Biotechnol.* **19**, 1060–1065.

Baldwin, D. N.., Shepherd, B., Kraemer, P., Hall, M., Sycuro, L., and Saloma, N. R. Identification of *Helicobacter pylori* genes contributing to stomach colonization. *Infect. Immun.* In press.

Chan, K., Kim, C. C., and Falkow, S. (2005). Microarray-based detection of *Salmonella enterica* serovar *Typhimurium* transposon mutants that cannot survive in macrophages and mice. *Infect. Immun.* **73**, 5438–5449.

Chun, K. T., Edenberg, H. J., Kelley, M. R., and Goebl, M. G. (1997). Rapid amplification of uncharacterized transposon-tagged DNA sequences from genomic DNA. *Yeast* **13**, 233–240.

Cooley, L., Kelley, R., and Spradling, A. (1988). Insertional mutagenesis of the *Drosophila* genome with single P elements. *Science* **239**, 1121–1128.

Hensel, M., Shea, J. E., Gleeson, C., Jones, M. D., Dalton, E., and Holden, D. W. (1995). Simultaneous identification of bacterial virulence genes by negative selection. *Science* **269**, 400–403.

Judson, N., and Mekalanos, J. J. (2000). Transposon-based approaches to identify essential bacterial genes. *Trends Microbiol.* **8**, 521–526.

Kumar, A., Vidan, S., and Snyder, M. (2002). Insertional mutagenesis: Transposon-insertion libraries as mutagens in yeast. *Methods Enzymol.* **350**, 219–229.

Liu, Y. G., and Whittier, R. F. (1995). Thermal asymmetric interlaced PCR: Automatable amplification and sequencing of insert end fragments from P1 and YAC clones for chromosome walking. *Genomics* **25**, 674–681.

Liu, Y. G., Mitsukawa, N., and Whittier, R. F. (1993). Rapid sequencing of unpurified PCR products by thermal asymmetric PCR cycle sequencing using unlabeled sequencing primers. *Nucleic Acids Res.* **21**, 3333–3334.

Liu, Y. G., Mitsukawa, N., Oosumi, T., and Whittier, R. F. (1995). Efficient isolation and mapping of *Arabidopsis thaliana* T-DNA insert junctions by thermal asymmetric interlaced PCR. *Plant J.* **8**, 457–463.

Manoil, C. (2000). Tagging exported proteins using *Escherichia coli* alkaline phosphatase gene fusions. *Methods Enzymol.* **326**, 35–47.

Salama, N. R., Shepherd, B., and Falkow, S. (2004). Global transposon mutagenesis and essential gene analysis of *Helicobacter pylori*. *J. Bacteriol.* **186**, 7926–7935.

Sassetti, C., and Rubin, E. J. (2002). Genomic analyses of microbial virulence. *Curr. Opin. Microbiol.* **5**, 27–32.

Sassetti, C. M., Boyd, D. H., and Rubin, E. J. (2001). Comprehensive identification of conditionally essential genes in mycobacteria. *Proc. Natl. Acad. Sci. USA* **98**, 12712–12717.
Shea, J. E., Hensel, M., Gleeson, C., and Holden, D. W. (1996). Identification of a virulence locus encoding a second type III secretion system in *Salmonella typhimurium*. *Proc. Natl. Acad. Sci. USA* **93**, 2593–2597.
Sherlock, G., Hernandez-Boussard, T., Kasarskis, A., Binkley, G., Matese, J. C., Dwight, S. S., Kaloper, M., Weng, S., Jin, H., Ball, C. A., Eisen, M. B., Spellman, P. T., Brown, P. O., Botstein, D., and Cherry, J. M. (2001). The stanford microarray database. *Nucleic Acids Res.* **29**, 152–155.

[11] Screening Transposon Mutant Libraries Using Full-Genome Oligonucleotide Microarrays

By Kelly M. Winterberg and William S. Reznikoff

Abstract

The experimental details for a high-throughput microarray-based screening technique for both detecting and mapping Tn*5* insertion mutants in parallel within a library are presented. Following Tn*5* mutagenesis, viable mutants are pooled and grown competitively under selective conditions. Chromosomal DNA is then isolated from each mutant pool. Biotin-labeled run-off *in vitro* RNA transcripts, representing the neighboring chromosomal DNA for each insertion remaining in the population, are generated using T7 promoters located at the ends of the transposon. Custom-designed, whole-genome oligonucleotide microarrays are used to analyze the labeled RNA transcripts and to detect each mutant in the library. Microarray data comparisons for each growth condition allow the identification of mutants that failed to survive the imposed growth selection. In addition, due to the density of the microarrays the genomic locations of the individual transposon insertions within each library can be identified to within 50 base pairs. Details for the *in vivo* Tn*5* mutagenesis procedure, mutant library construction and competitive outgrowth, T7 *in vitro* transcription/labeling, and microarray data analysis will be provided.

Introduction

In 1995 sequencing of the first bacterial genome of *Haemophilus influenzae* was completed (Fleischmann *et al.*, 1995). Since then 344 microbial genomes have been fully sequenced and over 550 more are in various stages of completion (http://www.ncbi.nlm.nih.gov). On average 30 to 40% of the

METHODS IN ENZYMOLOGY, VOL. 421
0076-6879/07 $35.00
DOI: 10.1016/S0076-6879(06)21011-5

genes in any sequenced genome are characterized as having unknown or hypothetical functions (Judson and Mekalanos, 2000; Lehoux *et al.*, 2001). A portion of these genes have homologues to other genes with unknown functions, but the remainder of these genes has no similarity to any genes in the available databases. Although genomic sequence data provide the basic genetic information for a given organism, they do not tell us the functions of each gene, when or how the genes are regulated, or under what growth conditions each gene is required by the organism. To use genomic sequence information to answer these questions, the development of high-throughput molecular and genetic approaches is needed.

Classically, gene functions have been characterized one gene at a time using conventional genetic techniques such as transposon mutagenesis and phenotype and/or loss-of-function screening. Although somewhat labor intensive, these methods have been quite effective for identifying gene functions. Recently, new high-throughput screening techniques have also been developed. Some of these techniques that have been used for studying phenotypes of transposon mutants in parallel have included *in vivo* expression technology (IVET) (Slauch *et al.*, 1994), signature-tagged mutagenesis (STM) (Hensel *et al.*, 1995), size-marker integration technique (SMIT) (Benton *et al.*, 2004), genetic footprinting (Smith *et al.*, 1995), transposon site hybridization (TraSH) (Sassetti *et al.*, 2001), and similar methods (Badarinarayana *et al.*, 2001; Winterberg *et al.*, 2005). Collectively, these high-throughput approaches have the potential to address the fitness of individual unknown bacterial mutants in a pool under certain conditions. This provides an initial starting point for choosing interesting genes to study individually in greater detail.

Advances in DNA microarray technology have also made it possible to not only measure changes in gene expression patterns for a given strain but to also perform comparative genomic hybridizations, genome resequencing, and DNA binding site analysis on a genome-wide scale. Two of the high-throughput methods mentioned above (Badarinarayana *et al.*, 2001; Sassetti *et al.*, 2001) use spotted DNA microarrays to analyze transposon mutant pools grown under different defined growth conditions. Following a PCR amplification of chromosomal DNA flanking each transposon insertion contained in the library, samples are hybridized to spotted DNA microarrays. The features/probes that were spotted onto the microarrays in these studies were either complete (Sassetti *et al.*, 2001) or partial (Badarinarayana *et al.*, 2001) open reading frames (ORFs). In both cases the microarray probes represented only a portion of the genome, and therefore were limited in detecting transposon insertions throughout the genome.

Here we present the experimental details for a technique that was developed to monitor transposon insertion mutants in parallel using custom-designed, high-density, whole-genome oligonucleotide microarrays (Winterberg *et al.*, 2005). In this chapter, we will describe this technique as it applies to the generation and screening of Tn*5* insertion libraries in *Escherichia coli* K-12. The principle and experimental details for this technique will be discussed, including Tn*5 in vivo* transposon mutagenesis, transposon mutant library construction, competitive outgrowth of transposon libraries, whole-genome oligonucleotide microarray design, mapping genomic locations of transposon insertions, and microarray data analysis methods. We will also indicate what types of modifications can be made and which parts of the protocols should be further optimized for application to other organisms.

Principle

This technique was developed to allow high-throughput parallel screening of transposon mutants under various *in vivo* and *in vitro* growth conditions using whole-genome oligonucleotide microarrays. Following transposon mutagenesis, libraries of viable mutants are constructed and competitively grown in both a control and a test condition (Fig. 1). The composition of each mutant library is analyzed by first isolating chromosomal DNA and digesting it to 1- to 1.5-kb fragments using an appropriate restriction enzyme. T7 promoters located at the ends of the transposon are used to generate in parallel biotin-labeled runoff *in vitro* transcripts of the chromosomal DNA flanking each transposon insertion contained in the library (Fig. 2). The labeled *in vitro* transcripts are hybridized to whole-genome oligonucleotide microarrays to identify which transposon insertions (and thus which mutants) were present in the original library (control condition), and which mutants have been lost from the library following competitive outgrowth in the test condition. In each case, the loss or out-competition of a mutant is inferred from the loss of *in vitro* transcript signal from a given transposon insertion. By using high-density, whole-genome oligonucleotide microarrays, the genomic locations of each transposon insertion for each mutant screened can be identified to within 50 bp of the insertion site. Using this technique, genes essential for growth in a variety of conditions can be easily identified.

The two key features that allow this technique to be employed are the T7 promoters contained within the transposon and the whole-genome oligonucleotide microarrays. Outward-facing T7 promoters are cloned into the 5′ and 3′ ends of the transposon prior to mutagenesis. These promoters are designed in a way that does not interfere with the transposition process, that is, the transposase binding sites should be left intact, but should allow run-off *in vitro* transcripts of the neighboring chromosomal DNA to be

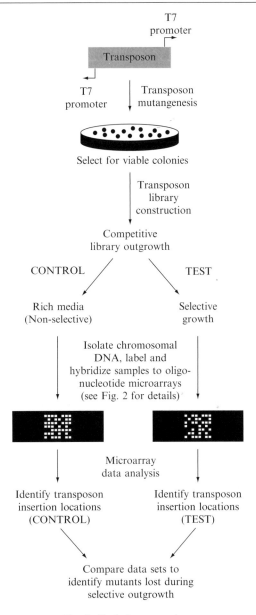

FIG. 1. Technique overview.

generated. Because the T7 transcripts are generated in a 5′ to 3′ direction, the transcripts generated from the left and right sides of the transposon will be homologous to the top (5′ to 3′) and bottom (3′ to 5′) strands of the

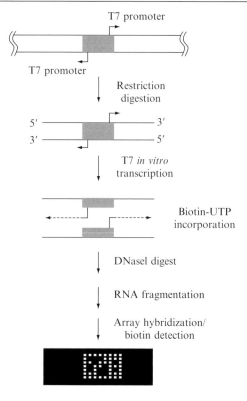

FIG. 2. *In vitro* transcription/biotin labeling schematic. Adapted and used with permission from the American Society for Microbiology.

neighboring DNA, respectively. When these transcripts are hybridized to whole-genome microarrays, the junction between the left- and right-side transcripts can easily be identified.

Second, whole-genome oligonucleotide microarrays are used to determine if a mutant is present in the library, and more specifically, where the transposon insertion within that mutant is located. The oligonucleotide probes representing the microarray are designed to represent both the top and bottom strands (5' to 3' and 3' to 5') of a given organism's genome, and are not specific to open reading frames or intergenic regions within the genome. This design strategy allows transposon insertions to be detected regardless of their location within the genome (intragenic or extragenic). Furthermore, the *in vitro* transcripts generated from each side of a given transposon insertion can be detected and depending on how closely spaced the oligonucleotide probes are located within the genome, transposon insertions can be mapped to within several base pairs of their genomic location. Many currently available commercial microarrays are designed specifically for gene expression studies and are thus limited in their representation of an organism's total genomic DNA. In the original development of this technique (Winterberg *et al.*, 2005), commercial microarrays using both oligonucleotide probes and spotted open reading frames were tested. In both cases, the design of the microarrays, although suitable for gene expression, failed to allow the detection of more than 50% of the insertions within a given transposon mutant library. When a custom-designed, whole-genome oligonucleotide microarray was implemented, each insertion in the library could easily be detected, suggesting that using whole-genome microarrays for tracking transposon insertion mutants was more appropriate.

Transposon Mutagenesis Protocol

In general, transposon mutagenesis is performed using a transposon containing outward-facing T7 promoters cloned into the 5′ and 3′ ends of the transposon and a selectable marker. In the original development of this technique (Winterberg *et al.*, 2005), *in vivo* Tn5 mutagenesis (Goryshin *et al.*, 2000) was used to deliver a modified transposon to competent *E. coli* cells by electroporation. The pMOD2 transposon-cloning vector (Epicentre Inc., Madison, WI) was used to clone two T7 promoters and a kanamycin-resistance gene into the Tn5 transposon. Transposome complexes were formed between the transposon DNA and purified hyperactive EK/LP Tn5 transposase (Bhasin *et al.*, 1999). Following electroporation and a short recovery period, kanamycin-resistant colonies were selected on agar plates. Approximately 10^5 kanamycin-resistant colonies per milliliter of transformed culture can typically be generated using this method.

Other transposon mutagenesis strategies can also be used with this method as long as the transposon contains a selectable marker and outward-facing T7 promoters. Care should be taken to choose a transposon that transposes fairly randomly to allow more complete coverage of the genome and to choose mutagenesis conditions that will minimize the occurrence of multiple transposon insertions within a single mutant. Southern blot analysis of several of the resulting transposon mutants can be used to verify that the mutagenesis resulted in single random insertions.

Tn5 Transposome Complexes

1. Perform a plasmid purification of the pMOD2 plasmid vector containing the transposon of interest. Alternatively, polymerase chain reaction (PCR) can be performed to amplify the transposon from the plasmid vector.

2. Digest approximately 10 μg of plasmid DNA or PCR product with *Psh*AI to release the transposon from the surrounding DNA.

3. Separate the transposon from the plasmid backbone using agarose gel electrophoresis. Purify the DNA from the transposon DNA band using either a gel extraction kit from Qiagen or other appropriate kit. (If PCR was used to generate the transposon DNA, a Qiagen PCR cleanup is sufficient to remove the digested ends away from the transposon DNA). Elute the DNA in deionized water.

4. Check the DNA concentration and purity by measuring the A_{260} and A_{260}/A_{280} ratio.

5. On ice, set up transposome complex reactions containing: 0.5 μg pre-cut transposon DNA, 4 μl of 10× binding buffer (10× concentration: 250 mM of Tris-acetate, pH 7.5, 1 M of potassium glutamate), and deionized water to 36 μl. Add different amounts of EK/LP hyperactive transposase ranging from 0.25 μg to 1 μg to each reaction. (The volume of transposase added to the reaction should never exceed 10% of the final reaction volume.) Deionized water should be added to each reaction to make the final volume 40 μl. A no-transposase control reaction should be included each time new transposome complexes are formed. (Optimization may be needed to determine the proper amounts of transposase to be used. Although hyperactive Tn5 transposase is commercially available [Epicentre Inc. Madison, WI], Tn5 transposase purified in the lab was used for this method.)

6. Incubate reactions at 37° for 2 hr.

7. Prior to electroporation, check the transposome complexes by running 4 μl of each reaction along with transposon DNA alone on a 1% agarose gel. The transposon DNA-only lane will provide a guide for where the transposon band within the transposome complex reactions should migrate. Transposome complexes will appear as a ladder of bands migrating slower than the transposon alone. The transposon band will appear diminished, concurrent with an increase in transposome complex bands.

8. Buffer exchange is used to decrease the amount of salt in each reaction prior to electroporation. This is done by spotting samples onto 0.05-μm filter discs floating in either 10% glycerol/5 mM of Tris-acetate (pH 7.5) or deionized water. This will help minimize arcing when the complexes are electroporated into electrocompetent cells. Transposome complexes exchanged into 10% glycerol/5 mM of Tris-acetate (pH 7.5) can be stored at –20° for up to 1 year.

In Vivo Tn5 *Mutagenesis*

Growth conditions are given for *E. coli* and should be changed as needed for other organisms.

1. Make transposome complexes with the transposon of interest and hyperactive EK/LP Tn5 transposase (as described above).

2. Using standard electroporation conditions, electroporate 1 to 2 μl of complexes in binding buffer into 50 μl of electrocompetent cells. If complex reactions were buffer exchanged into 10% glycerol/5 mM of Tris-acetate (pH 7.5), up to 10 μl can be electroporated. If water was used for buffer exchange, presumably the entire binding reaction can be electroporated into the cells. This may be needed if the competency of the recipient strain is low.

3. Recover cells in 1 ml of Luria-Bertani (LB) medium at 37° for 1 hr with shaking (250 rpm). Incubations of more than 1 hr can lead to the accumulation of sibling insertions that are most likely due to the outgrowth/doubling of the recovering cells.

4. Isolate transformants by plating 50 to 100 μl of cells per LB plate containing the appropriate antibiotic. Incubate at 37° overnight. Typically 10^5 individual transposon mutants can be obtained from 1 ml of electro-porated culture, but this will depend on the competency of the recipient cells and the amount of transposome complexes that were electroporated.

5. Perform Southern blot analysis of several selected mutants to verify that random single insertions were isolated in the mutagenesis.

Library Construction Protocol

Following transposon mutagenesis, mutant libraries can either be made directly from the selection plates or by first stocking individual mutants in 96-well format and then combining individual mutants. For the former method colonies should be dislodged and scraped together with ~2 to 3 ml of LB broth (per plate) using a glass spreading rod. The size of the library can be increased by combining the colonies from several plates. Aliquots of this mixture can be stored in 15% glycerol at –80°.

Mutant libraries can also be constructed by first picking individual mutants into separate wells of a polystyrene 96-well plate containing LB broth (~200 μl per well). Incubate 96-well plates at 37° in a stationary incubator overnight. Stock each plate of mutants in 15% glycerol (final concentration) by transferring the overnight culture by multichannel pipe-tor into a sterile polypropylene 96-well plate containing the appropriate amount of glycerol. Store plates at –80°. Libraries of various sizes can be constructed from these 96-well stock plates. To do this, mutants should be inoculated from the frozen stock plates into fresh LB broth in 96-well format using a 48- or 96-well replica pinner transfer device. Following overnight growth at 37°, combine equal volumes of cell culture in a sterile reservoir or tube to create libraries of various sizes. A few individual OD_{600} culture readings should be taken to verify that the cultures are at similar densities prior to pooling. Aliquots of this library can be stored in 15% glycerol at –80°. Although this method is a bit more labor intensive, the number of mutants being combined into a library can be controlled easier and it provides a stock of individual transposon mutants that can be used for additional experiments.

In some cases, the size limit of the library may need to be determined empirically. In the development of this technique, mutant libraries containing 94 mutants allowed scoring of 100% of the members. Larger libraries containing 188, 376, and 564 mutants permitted the detection of only 79.8%, 72.9%, and 59% of the library members, respectively (Winterberg et al., 2005). This was most likely due to the dilution of T7 promoter sequences as more mutant chromosomal DNA is added to the population. Modifications to enhance the T7 promoter containing DNA fragments in the template population prior to the in vitro transcription labeling procedure could be added. This might include modifying the stated procedure to include a separation step where all DNA fragments containing a T7 promoter and transposon end sequence could be separated away from the "nonspecific" chromosomal DNA. This could be done by running the digested DNA over a column that would specifically bind the T7 and/or transposon end sequences. Additionally, the competitive outgrowth test condition may also restrict the size of the mutant library. For example, some in vivo animal screening models have a limit to the complexity of the mutant library that can successfully establish an infection and be screened in a single animal (Bahrani-Mougeot et al., 2002). Therefore, the experimental design may affect the complexity of the transposon library that can be screened effectively using this method.

Competitive Outgrowth of Mutant Libraries Protocol

Competitive outgrowth of mutant libraries is used to screen many transposon mutants in parallel to identify mutants that cannot survive an imposed (test) growth condition as compared to a control condition. Mutants from the test condition that fail to grow or are out-competed by the other mutants in the library are identified by microarray analysis (see below). The control condition should be a nonselective rich media that preferably is similar to the conditions originally used for the selection of transposon insertion mutants. This control condition should permit all of the transposon insertion mutants to grow. A test condition such as growth in minimal media, high/low pH, heat shock, and in vivo growth in animals, plants, and/or insects should be chosen to identify specific classes of mutants. Additionally, as a reverse approach, mutant pools that are specific for certain pathways or cellular functions presumably could be used with this technique to determine the composition and/or nutrient make-up of an unknown environmental growth condition.

The following procedure can be used for performing competitive outgrowth in rich and minimal media to identify M9 minimal-glucose auxotrophs.

1. Thaw a transposon mutant library aliquot on ice and dilute it 1/10 in LB broth to ~4.5 ml. Incubate for 1 to 3 hr at 37° with shaking (250 rpm). (This recovery of the frozen library introduces a short period of competitive growth and possible cell doubling, and should therefore be minimized to prevent the loss of any mutants within the population.) Alternatively, if mutants were stocked individually, frozen 96-well plates should be used to inoculate 96-well plates containing fresh media. Grow mutants overnight at 37°, pool them in a sterile reagent reservoir, and mix well.
2. Pellet the cells by centrifugation and discard the supernatant.
3. Wash the pellet three times with ~4.5 ml of phosphate-buffered saline, pelleting the cells after each resuspension.
4. Resuspend the final cell pellet in 4.5 ml of fresh phosphate-buffered saline.
5. Dilute the resuspended library cells 1:50 into ~100 ml each of both rich (control condition) and M9 minimal media supplemented with 0.2% glucose (test condition).
6. Incubate the cultures at 37° with shaking (250 rpm).
7. Take OD_{600} readings to determine doubling times/generations of each culture.
8. When the cells have reached late log phase/early stationary phase, harvest the cells in each condition by pelleting the cells in a tabletop centrifuge.
9. Harvest chromosomal DNA separately from each cell pellet in preparation for *in vitro* transcription and biotin labeling.

Mutant Labeling Protocol

Chromosomal DNA isolated from competitively grown mutant libraries is prepared for *in vitro* transcription/biotin labeling by first digesting it with a restriction enzyme that leaves blunt ends and will result in average DNA fragment sizes of ~1 to 1.5 kb. Run-off *in vitro* transcription is used to generate biotin-labeled RNA transcripts of the DNA directly flanking each transposon insertion still remaining in the mutant library following competitive outgrowth.

DNA Template Preparation

1. Quantify and check the quality of the isolated DNA by measuring the A_{260}/A_{280} ratio. The A_{260}/A_{280} reading should be ~1.8 to 1.9.

2. Digest 40 to 50 μg of DNA to completion using an enzyme that leaves blunt ends. (A blunt-end cutter is used because overhangs of DNA can be used as nonspecific templates by T7 polymerase.) Typically, *Hin*cII or *Fsp*I is used for digestion of *E. coli* strains resulting in average fragment sizes ranging from ~1 to 1.5 kb in length.

3. Add an equal volume of phenol chloroform isoamyl alcohol to the digested DNA and vortex mix for ~30 sec. Transfer mix to phase-lock gel tubes (Eppendorf) for easier phase separation. Spin tubes at 13,000 rpm (~17,900 × g) in a conventional table-top microcentrifuge for ~1 min to separate the phases. Transfer the aqueous phase to an RNase-free 1.5-ml Eppendorf tube.

4. Ethanol precipitate the digested DNA by adding 1/10 volume of RNase-free, 3 M sodium acetate (NaOAc), pH 5.2, and 2–1/2 volumes of cold 95% ethanol (RNase-free). Invert the tube 20 to 30 times until well mixed. Precipitate at –20° for at least 2 to 3 hr (preferably overnight). Spin samples at 13,000 rpm (~17,900 × g) in a conventional table-top micro-centrifuge for 10 min at 4°. Wash pellets once with 1 ml of RNase-free 70% ethanol. Allow pellets to dry for 10 to 15 min at room temperature.

5. Resuspend the DNA pellet to a concentration of ~2 μg/μl in diethyl-pyrocarbonate (DEPC)-treated/nuclease-free water. (Typically, this takes ~10 to 15 μl of DEPC-treated water for each 40 to 50 μg of digest). Measure the A_{260} and A_{260}/A_{280} ratio following resuspension of the DNA pellet.

In Vitro *Transcription Reaction*

This step uses the Epicentre Ampliscribe T7 High Yield *in vitro* tran-scription kit to generate biotin-labeled RNA transcripts from the ends of each transposon insertion within the population. Other *in vitro* transcription kits should also work at this step.

1. Mix the following at *room temperature*:
     3 μl of 10X T7 reaction buffer
     10 μl of 10 mM Biotin-16-UTP (Enzo)
     1.5 μl of 100 mM ATP
     1.5 μl of 100 mM GTP
     1.5 μl of 100 mM CTP
     0.5 μl of 100-mM UTP
     2.5 μl of 100-mM DTT
     10 μg of digested, precipitated DNA (in a volume of less than 6.5 μl)
     Add DEPC-treated water as needed to 27 μl
2. Add 3 μl of T7 RNA polymerase (from –20°) and mix well. (Total volume should now be 30 μl.)

3. Incubate at $37°$ overnight in a heat block. (Shorter times may yield less RNA, but may still be sufficient.)

4. Add 2 μl of DNaseI and incubate for 15 min at $37°$. This removes all chromosomal DNA that may compete with the labeled RNA during microarray hybridization.

5. To clean up the labeled samples prior to microarray hybridization, add 470 μl of DEPC-treated water to each 32 μl sample, transfer to a YM10 microcon column, and spin as recommended until $\sim$20 μl remain. Recover the retentate in a fresh RNase-free collection tube.

6. Quantify the RNA by measuring the A_{260}, and check the purity by measuring the A_{260}/A_{280} ratio. Typically, this is done on a NanoDrop spectrophotometer to minimize sample loss.

Microarray Design and Hybridization Protocol

A significant aspect of this technique is to use a microarray that represents most of the genomic DNA of an organism. During the optimization of this technique, it was determined that microarrays designed specifically for gene expression analysis (both spotted arrays and oligonucleotide arrays, and specifically, Affymetrix GeneChip arrays) failed to detect over 50% of the insertions from a transposon insertion library containing precharacterized transposon insertion mutants (Winterberg et al., 2005). For this reason, care should be taken to design and/or use microarrays that provide near-complete genomic coverage of the strain or chromosomal regions being investigated.

1. Prior to hybridization, RNA samples should be fragmented. To do this, mix DEPC-treated water, 5 to 10 μg of RNA, and 5X fragmentation buffer (200 mM of Tris-acetate, pH 8.1, 500 mM of potassium acetate, and 150 mM of magnesium acetate) to a final volume of 30 μl. Fragmentation buffer should be used at a final concentration of 1X. Incubate for 10 min at $95°$. Hold on ice until prehybridization and hybridization mixtures are prepared.

2. Prehybridization and hybridization to whole-genome oligonucleotide microarrays or other suitable microarray should be performed as recommended by the manufacturer.

3. Following microarray hybridization, biotin-labeled transcripts hybridized to oligonucleotide microarrays are detected by first staining the microarrays with a Cy3-streptavidin conjugated fluorophore dye, and then scanning with a confocal scanning laser.

4. Signal intensities of each oligonucleotide probe on the microarray are extracted from the scanned microarray image based on the X and Y positions on the microarray.

Data Analysis

Although many microarray data analysis programs exist, it is important to use a program that allows analysis of the individual probe signal intensities, rather than a program that calculates the average probe signal intensity for each individual gene, as is often the case for gene expression analysis software programs. Most microarray data sets will contain over 65,000 records, and therefore cannot be analyzed using a spreadsheet program such as Microsoft Excel. Database programs like Microsoft Access or other custom programs written in PERL and/or MySQL can be used to organize and sort the microarray data.

Probe signal intensities are extracted from each microarray based on X and Y coordinates from the microarray. The signal intensity data and genome position for each probe can be linked together using the original microarray library file that describes each probe's X and Y position on the microarray and the DNA sequence and genome position for each probe. When the genome position and signal intensity data for each oligonucleotide probe have been linked, the microarray data can be sorted in genome order. Both strands of DNA can be analyzed together, but a common factor of (-1) should be multiplied to each probe signal intensity for one of the strands (typically the $3'$ to $5'$ or bottom strand). This will allow differentiation between probes that are located at the same genome position but are homologous to opposite strands.

Software programs that allow several histograms of data to be plotted relative to each other are very effective for visualizing this type of microarray data. GenVision (DNASTAR) is an example of such a program, and was used during the development of this technique. Files representing (1) the gene names, (2) gene lengths and coding strand (sense or antisense), (3) microarray probe-signal intensities, and (4) restriction enzyme digestion sites all relative to the position in the genome can be constructed using gene annotation tables and the extracted microarray data. As shown in Figs. 3 and 4, GenVision allows the visualization and comparison of each of these histograms of information.

In the development of this technique (Winterberg *et al.*, 2005), it was determined that mutants diluted 10^{-1} in a library of 50 total mutants will show a noticeable signal intensity decrease, and mutants diluted 10^{-3} will no longer be detected on the microarray. Thus, the limit of detection for this technique is between 10^{-1} and 10^{-3}. A decrease in signal may also represent a region of the chromosome that T7 RNA polymerase does not transcribe well, but this decrease should not change between the two growth conditions being tested.

Array normalization and/or specific algorithms to pick out stretches of higher signal intensities may or may not be necessary depending on the

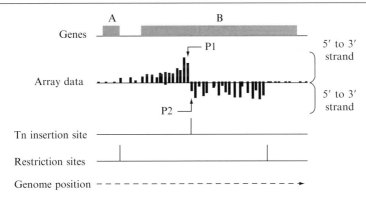

FIG. 3. Schematic for plotting and analyzing microarray data using GenVision (DNASTAR Inc.). Gene names (A or B) are shown in the top panel. Grey boxes representing the lengths and genome location of genes A and B, and are located above the median line (indicating that the genes are encoded on the sense strand of DNA). The second panel represents the microarray probe data shown as black bars extending above (5′ to 3′ strand) or below (3′ to 5′ strand) the median line. The width of the bar represents the length of the probe, and the height of the bar represents the probe signal intensity. Probe signal intensities are plotted above the median line for probes homologous to the sense or top strand of DNA and below the line for probes homologous to the antisense or bottom strand of DNA. This is done by multiplying (–1) to each signal intensity for the bottom strand probes. The third panel represents the predicted transposon insertion site. The junction between the divergent *in vitro* transcripts represents where the transposon originally inserted. The fourth panel represents the restriction enzyme cut sites within the genome for the region shown. P1, first top strand probe of the left side T7 *in vitro* transcript. P2, first bottom strand probe of the right side T7 *in vitro* transcript. Adapted and used with permission from the American Society of Microbiology.

specific application of the technique. In general, if the probe signal intensities are plotted in genome order with respect to the genes in the organism, mutants that were out-competed during selective outgrowth and the locations of each insertion in the library should be very clear. More specifically, the genomic locations of the transposon insertions can be mapped using the microarray data (Fig. 3). The junction between the probes representing the left and right side *in vitro* transcripts provides a narrow range (~50 bp if the probes are spaced every 50 bp) for the location of the original transposon insertion. This genomic range can be found by subtracting the genome positions of the first bottom-strand probe of a right-side T7 *in vitro* transcript (P2 in Fig. 3) and the first top-strand probe for the corresponding left-side T7 *in vitro* transcript (P1 in Fig. 3). The genomic spacing of the microarray probes and the visualization of the signal intensity data relative to the genes in the genome allows immediate gene identification for each mutant within the transposon library. One perceived drawback of this method is that it may be difficult to isolate the original transposon insertion

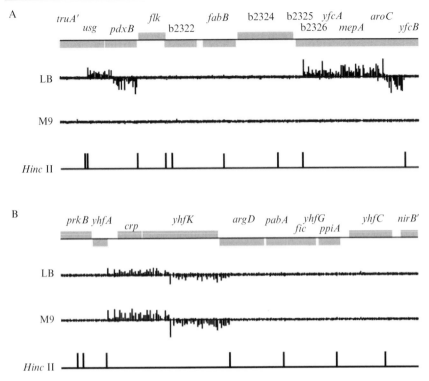

Fig. 4. Sample microarray data for transposon mutant libraries grown in rich (LB) and minimal (M9) media. Two separate regions of the *Escherichia coli* chromosome are depicted. (A) Insertions in *pdxB* and *aroC* can be detected in the LB grown cells but not in the M9 minimal cells, confirming the M9 minimal auxotrophy of mutants lacking either of these genes. (B) The insertion in *yhfK* can be detected in both the LB and M9 minimal grown cells, indicating that *yhfK* is not required for survival in M9 minimal media. Adapted and used with permission from the American Society for Microbiology.

mutant following competitive outgrowth and array analysis. But given that this method provides immediate gene identification for every mutant screened, knockout mutant construction for genes of interest and follow-up studies can begin immediately following the analysis of the microarray. This eliminates the need to isolate and sequence the transposon insertion site from the original mutant in the library.

Conclusion

This technique provides a high-throughput method for screening libraries of transposon insertion mutants under various growth conditions. The use of high-density whole-genome oligonucleotide microarrays affords a distinct advantage over other similar methods in that the genomic location for each transposon insertion within the library can be immediately identified and used to pinpoint the disrupted gene/chromosomal element. Individual knockout mutants can then be constructed to verify the growth phenotypes (outcompetition of a given mutant) depicted by the microarray results. The increase in available DNA sequence data in combination with the advancements in DNA microarray technology and transposomic tools should allow this method to be tailored to many different biological systems.

Acknowledgment

Research has been funded by the National Science Foundation (MCB 0315788).

References

Badarinarayana, V., Estep, P. W., 3rd, Shendure, J., Edwards, J., Tavazoie, S., Lam, F., and Church, G. M. (2001). Selection analyses of insertional mutants using subgenic-resolution arrays. *Nat. Biotechnol.* **19,** 1060–1065.

Bahrani-Mougeot, F. K., Buckles, E. L., Lockatell, C. V., Hebel, J. R., Johnson, D. E., Tang, C. M., and Donnenberg, M. S. (2002). Type 1 fimbriae and extracellular polysaccharides are preeminent uropathogenic *Escherichia coli* virulence determinants in the murine urinary tract. *Mol. Microbiol.* **45,** 1079–1093.

Benton, B. M., Zhang, J. P., Bond, S., Pope, C., Christian, T., Lee, L., Winterberg, K. M., Schmid, M. B., and Buysse, J. M. (2004). Large-scale identification of genes required for full virulence of *Staphylococcus aureus. J. Bacteriol.* **186,** 8478–8489.

Bhasin, A., Goryshin, I. Y., and Reznikoff, W. S. (1999). Hairpin formation in Tn5 transposition. *J. Biol. Chem.* **274,** 37021–37029.

Fleischmann, R. D., Adams, M. D., White, O., Clayton, R. A., Kirkness, E. F., Kerlavage, A. R., Bult, C. J., Tomb, J. F., Dougherty, B. A., and Merrick, J. M. (1995). Whole-genome random sequencing and assembly of *Haemophilus influenzae* Rd. *Science* **269,** 496–512.

Goryshin, I. Y., Jendrisak, J., Hoffman, L. M., Meis, R., and Reznikoff, W. S. (2000). Insertional transposon mutagenesis by electroporation of released Tn5 transposition complexes. *Nat. Biotechnol.* **18,** 97–100.

Hensel, M., Shea, J. E., Gleeson, C., Jones, M. D., Dalton, E., and Holden, D. W. (1995). Simultaneous identification of bacterial virulence genes by negative selection. *Science* **269,** 400–403.

Judson, N., and Mekalanos, J. J. (2000). Transposon-based approaches to identify essential bacterial genes. *Trends Microbiol.* **8,** 521–526.

Lehoux, D. E., Sanschagrin, F., and Levesque, R. C. (2001). Discovering essential and infection-related genes. *Curr. Opin. Microbiol.* **4,** 515–519.

Sassetti, C. M., Boyd, D. H., and Rubin, E. J. (2001). Comprehensive identification of conditionally essential genes in mycobacteria. *Proc. Natl. Acad. Sci. USA* **98,** 12712–12717.

Slauch, J. M., Mahan, M. J., and Mekalanos, J. J. (1994). *In vivo* expression technology for selection of bacterial genes specifically induced in host tissues. *Methods Enzymol.* **235,** 481–492.

Smith, V., Botstein, D., and Brown, P. O. (1995). Genetic footprinting: a genomic strategy for determining a gene's function given its sequence. *Proc. Natl. Acad. Sci. USA* **92,** 6479–6483.

Winterberg, K. M., Luecke, J., Bruegl, A. S., and Reznikoff, W. S. (2005). Phenotypic screening of *Escherichia coli* K-12 Tn5 insertion libraries, using whole-genome oligonucleotide microarrays. *Appl. Environ. Microbiol.* **71,** 451–459.

[12] Creating Recombination-Activated Genes and Sequence-Defined Mutant Libraries Using Transposons

By LARRY GALLAGHER, CHERI TURNER, ELIZABETH RAMAGE, and COLIN MANOIL

Abstract

The properties of a collection of transposon Tn5 derivatives that generate reporter gene fusions and internal protein tags are summarized. Procedures utilizing several of the transposons for generating genes activated by Cre-loxP recombination and for creating large sequence-defined mutant libraries are described in detail.

Introduction

The broad utility of transposable elements as tools for genetic and genomic analysis is well-established (Hayes, 2003). The mutations they generate are precisely defined, are easily sequence mapped, are usually limited to one event per mutated genome or plasmid, and generally create strong loss-of-function alleles. Most useful transposable elements carry selectable markers, readily allowing mutated chromosomal genes to be cloned or transferred genetically between strains. In this article, we first summarize the properties of transposon Tn5 derivatives we have developed. We then describe recently developed applications employing several of the transposons either to create genes whose functions are controlled by site-specific recombination or to generate comprehensive sequence-defined mutant libraries.

Transposon Tn5 Derivatives

Transposons incorporating a variety of useful genetic elements have been engineered. These include reporter genes that generate transcriptional or translational gene fusions, unique signature tags to allow individual mutants to be tracked in pools (Mecsas, 2002; Shea et al., 2000), and sequences acted on by site-specific recombinases to allow the generation of internal protein tags (Bailey and Manoil, 2002; Manoil, 2000; Manoil and Traxler, 2000).

The characteristics of many of the transposable elements constructed in our laboratory are summarized in Table I, Fig. 1, and at www.gs.washington. edu/labs/manoil/transposons.htm. A simplified nomenclature for the transposons (e.g., T1-T22) is introduced. The full sequences and construction

METHODS IN ENZYMOLOGY, VOL. 421
0076-6879/07 $35.00
DOI: 10.1016/S0076-6879(06)21012-7

histories of the elements are available through the website, and as other transposons are constructed, they will be added. Several of the key attributes of transposons in the collection are summarized below.

Reporter Fusions

Many of the transposons can generate β-galactosidase or alkaline phosphatase gene fusions (Manoil *et al.*, 1990; Silhavy, 2000; Silhavy and Beckwith, 1985). Elements creating translational or transcriptional *lacZ* fusions are represented.

Recombination and Gene-Tagging Features

Several of the transposons were designed to allow them to be converted into short in-frame insertions in the target gene using additional *in vitro* or *in vivo* steps. The insertions range from 31 to 63 codons and most encode epitopes and/or hexahistidine metal–affinity purification tags. The *in vitro* method for generating the in-frame insertions utilizes transposons (Tn*lacZ*/in or Tn*phoA*/in) carrying BamH1 restriction sites positioned near the ends of the elements. After insertion of the transposon into a plasmid carrying the target gene of interest (and lacking BamH1 sites), BamHI cleavage and ligase joining are used to convert the transposon insertion into a short in-frame insertion. The *in vivo* method for generating the internal insertions utilizes transposons carrying *loxP* or *FRT* sites at their ends. Exposure of mutagenized chromosomal or plasmid genes to the appropriate recombinase deletes sequences between the recombination sites and leaves a short in-frame insertion. Since the recombination events remove the transposon antibiotic-resistance determinants, the resistance markers may be reutilized in generating multiple mutants ("marker recycling"). Most of the transposons used for internal tagging also generate *lacZ* or *phoA* translational fusions. By selecting insertions that generate active reporter fusions, it is possible to selectively isolate insertions in the appropriate orientation and reading frame for conversion into the in-frame internal tags.

In the special case that an internal tag generated by site-specific recombination does not eliminate function of a gene, expression of the recombinase activates the gene. This activation has been used as the basis of sensitive, irreversibly activated, whole-cell biosensors (Turner, unpublished). A protocol for screening for recombination-activated alleles of reporter genes is provided below.

Delivery

One of two delivery methods is used for mutagenesis by most of the transposons. The first utilizes suicide plasmids delivered by conjugation

TABLE I
TRANSPOSABLE ELEMENTS

	Transposable element[a]	Vector[d]	Delivery method[g]	Resistance marker[j]	Reporter element[k]	Deletion method	Internal tag[m]	Tn-specific primers[n]	CEKG-2 primers[p]	References
	<KAN-2>	(none)	Tpm	kan	none	none	none	K	C,D,E	Epicentre[q]
T1	TnphoA	phage λ TnphoA	Transduction	kan	phoA TL	none	none	H	A,B,C	Manoil and Beckwith, 1985
T2	TnlacZ	phage λ TnlacZ	Transduction	kan	lacZ TL	none	none	L2	A,B,C	Manoil, 1990
T3	TnphoA/in	phage λ TnphoA/in	Transduction	cm	phoA TL	BamHI/ligase	31-codon	H	A,B,C	Manoil and Bailey, 1997
T4	TnlacZ/in	phage λ TnlacZ/in	Transduction	cm	lacZ TL	BamHI/ligase	31-codon	L2	A,B,C	Manoil and Bailey, 1997
T5	ISphoA/hah-cm	pCM638, pCM665[e]	Conj	cm	phoA TL	loxP	63-codon: HA, H_6	H	A,B,C	Bailey and Manoil, 2002
T6	ISphoA/hah-tc	pCM639	Conj	tc	phoA TL	loxP	63-codon: HA, H_6	H	A,B,C	Jacobs et al., 2003
T7	ISlacZ/hah-cm	pIT1	Conj	cm	lacZ TL	loxP	63-codon: HA, H_6	L	A,B,C	This paper[r]
T8	ISlacZ/hah-tc	pIT2	Conj	tc	lacZ TL	loxP	63-codon: HA, H_6	L	A,B,C	Jacobs et al., 2003
T9	ISΩ/hah	pCM1008	Tpm	strep/spec	none	loxP	54-codon: HA, H_6	CT1, CT28[o]	N/D	This paper
T10	IScm/FRT	pCM1767	Tpm	cm	none	FRT	35-codon	CT24[o]	N/D	This paper
T11	ISlacZY/hah-cm	pLG33	Conj	cm	lacZY TS[l]	loxP	63-codon: HA, H_6	L2	A,B,C	This paper
T12	ISlacZY	pLG42, pLG43[f], pLG44[f], pLG49	Tpm[h] or Tpm (PshAI)[i]	cm	lacZY TS	none	none	L2	A,B,C	This paper
T13	mTn5*-lacZ1-kan	pLG48b	Tpm (PshAI)	kan	lacZ TS	none	none	L2	C,D,I	This paper
T14	mTn5*-lacZ1-em	pLG51	Tpm (PshAI)	erm	lacZ TS	none	none	L2	C,D,I	This paper
T15	ISR6K-em[b]	pLG52a, pLG53, pLG55a	Tpm (PshAI)	erm	none	none	none	E	A,C,E	This paper
T16	ISR6K-kan[b]	pLG56a	Tpm (PshAI)	kan	none	none	none	N	B,E,F	This paper
T17	ISFn1[c]	pLG61a	Tpm	kan[Fn]	none	none	none	K or F1	C,D,E	This paper
T18	ISFn2[c]	pLG62a	Tpm	kan[Fn]	none	none	none	K or F2	C,D,E or B,D,E	This paper
T19	ISFn1/FRT[c]	pLG65a	Tpm	kan[Fn]	none	FRT	none	K or F1	C,D,E	This paper
T20	ISFn2/FRT[c]	pLG66a	Tpm	kan[Fn]	none	FRT	none	K or F2	C,D,E	This paper
T21	ISgfp-Fn2/FRT[c]	pLG67	Tpm	kan[Fn]	gfp TS	FRT	none	K	C,D,E	This paper
T22	ISlacZ-Fn2/FRT[c]	pLG69	Tpm	kan[Fn]	lacZ TS	FRT	none	L2 or K	A,B,C or C,D,E	This paper

[a] Each transposable element is identified by both a unique "T" number and a transposon name.

[b] Transposon is derived from plasmid pMOD-3 (Epicentre) and carries an R6K origin of replication.

[c] In T17 and T19, an endogenous *F. novicida* promoter (for FTN-1451, the *F. novicida* orthologue of *F. tularensis* Schu4 gene omp26 [FTT1542c]) drives a kanamycin-resistance gene that retains its own translation initiation region. In T18, T20, T21, and T22, the *F. novicida* promoter drives a translational gene fusion between the native *F. novicida* gene and the kanamycin-resistance ORF.

[d] Multiple plasmids listed for a single transposon represent different construction histories. The transposons they carry are functionally equivalent, although minor sequence differences may exist within the transposon sequences. Other elements of the plasmids may also differ. Complete sequences and construction histories are available at www.qs.washington.edu/labs/manoil/transposons.htm.

[e] The transposase gene in pCM665 is a more active derivative of the one in pCM638 (Zhou *et al.*, 1998).

[f] pLG43 and pLG44 carry the R6K origin as their only replication origin.

[g] Tpm, transposon—transposase complex ("Transposome") transformation; Conj, conjugation; Tpm (PshAI), transposon can be precisely excised by PshAI digestion prior to transposome assembly.

[h] Transposon end sequences in pLG42, pLG42, pLG43, and pLG44, while functional, are not perfect matches to the hyperactive sequence needed for optimal transposome efficiency (Zhou *et al.*, 1998).

[i] The "Tpm (PshAI)" delivery method is a feature of pLG49, but not of pLG42, pLG43, or pLG44.

[j] kan, kanamycin resistance; tc, tetracycline; cm, chloramphenicol; strep/spec, streptomycin/spectinomycin; erm, erythromycin; kan^{Fn}, kanamycin driven by a *F. novicida* promoter.

[k] TL, translational fusion; TS, transcriptional fusion.

[l] In T11, the loxP site adjacent to the *lacZ* gene appears to contain promoter elements that are active in *E. coli*.

[m] For some transposons, insertions that are in the proper orientation and reading frame can be converted by the deletion method shown into small internal gene tags. For most tags, partial codons are present at both ends of the defined insertion sequences and are completed by the flanking nucleotides at the insert site. The number of codons shown for each tag includes those created by the 9-bp duplication produced by Tn5 transposition. Specific features encoded by some tags: HA, hemagglutinin epitope; H$_6$, hexahistidine.

[n] The set of three primers used for round 1 PCR, round 2 PCR and sequencing, respectively (see Fig. 3B and protocol in text), follow: K, primers kan2–211, kan2–145, and kan2–125; H, primers hah-166, hah-138, and hah-114; L, primers lacZ-211, lacZ-148, and lacZ-124L; L2, primers lacZ-211, lacZ-148, and lacZ-124L2; E, primers erm-204, erm-138, and erm-106; N, primers nptF-186, nptF-130, and nptF-105; F1, primers 806b-248, 806b-214, and 806-182; F2, primers 806c-208, 806-182, and 806–98. Primer sequences are available through the website.

[o] Three-step mapping with transposon-specific primers (see previous note) has not been done for T9 and T10. The individual primers listed may be used for one-step sequencing out of the transposons. Primer sequences are available through the website.

[p] For round 1 PCR, a mixture of three "CEKG-2" primers (semidegenerate primers with short defined sequences at their 3′ ends; see protocol in text) is used. Primers must be chosen whose defined 3′ ends (four or five nucleotides) do not anneal within the transposon between the transposon-specific primer site and the end of the transposon. Suggested mixtures of primers CEKG-2A, B, C, D, E, F, G, and H are listed. Primer sequences are available through the website. N/D, not determined.

[q] Epicentre Biotechnologies, www.epicentre.com.

[r] I. Thaipisuttikul, personal communication.

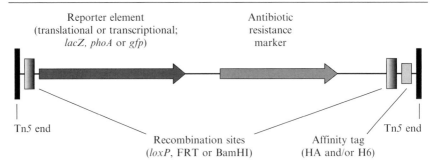

FIG. 1. Transposon features. All transposons carry Tn5 ends and an antibiotic-resistance marker. The Tn5 ends for some transposons are modified for optimal use with purified transposase. The other features shown are included in some of the transposons (Table I). HA, hemaglutinin epitope; H6, hexahistidine metal-affinity moiety.

based on pUT (Herrero *et al.*, 1990), incorporating the broad-host-range plasmid RK2 conjugation origin. The plasmids deliver both the transposon and the transposase gene to target cells, but the transposase gene is not included in the transposing unit.

An alternative delivery method, particularly useful for species that are not amenable to the suicide plasmid–based methodology, is based instead on electroporation of transposase–transposon complexes formed *in vitro* (Goryshin *et al.*, 2000). The assembly of such complexes requires slightly different 19-bp transposon end sequences than those found in wild-type Tn5 (Zhou *et al.*, 1998). Transposons with the modified ends may also be inserted into isolated DNA *in vitro*.

Recombination-Activated Alleles of Reporter Genes

We recently developed transposon-based methods to screen for derivatives of plasmid-borne genes that are activated by site-specific recombination. The procedure is based on the *in vitro* insertion of an element (such as ISΩ/hah) that can be acted on by Cre or FLP recombinase to generate a short in-frame insertion. If the original inactivating transposon insertion is at a site in the gene that tolerates such short insertions ("permissive" sites), the transposon insertion constitutes a recombination-activated version of the gene. We have used this method to construct recombination-activated derivatives of several reporter genes (*lacZ*, *luxCDABE*, *phoA*, and *gfp*).

Identifying Cre-Activated lacZ Derivatives Using ISΩ/hah

The protocol specifies the steps used to identify permissive sites in a *lacZ* gene carried in pBR322 (pMLB1101) using transposon ISΩ/hah (carried in plasmid pCM1008, a derivative of pUT [Herrero *et al.*, 1990]).

Analogous procedures were used for targeting the other reporter genes. Insertions of ISΩ/hah are converted into short (54-codon) insertions by the activity of Cre at the loxP sites (Fig. 2).

1. Mutagenize the target plasmid (e.g., pMLB1101) with ISΩ/hah *in vitro*. Incubate the target plasmid and pCM1008 (rendered linear by treatment with XhoI) with purified Tn5 transposase following the supplier's instructions (Epicentre). Drop dialyze the mutagenesis reaction to remove salts before electroporation (Maloy *et al.*, 1996).

2. Electroporate 10 to 400 ng of DNA from the dialyzed reaction into a streptomycin-sensitive or spectinomycin-sensitive, *lacZ⁻* pir⁻ strain of *E. coli* such as CC118 (Manoil and Beckwith, 1985). After outgrowth in LB with aeration for 1 hr, plate dilutions on agar supplemented with streptomycin (100 μg/ml) or spectinomycin (50 μg/ml), X-gal (40 μg/ml), and IPTG (0.2 m*M*). Incubate overnight at 37°.

3. Replica print from plates with 100 to 200 colonies onto L agar supplemented with X-gal (40 μg/ml), IPTG (0.2 m*M*), and CaCl₂ (5 m*M*)

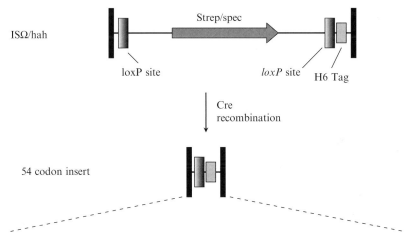

CTGTCTCTTATACACAACTgggatctgataacttcgtataatgtatgctatacgaagttattaattaagcatcaccatcaccatc actacccgtacgacgtgccggactacgcccgagatccagcaggttggcaAGATGTGTATAAGAGACAGnnnnnnnnnn

(LSLIHNWDLITSYNVCYTKLLIKHHHHHHYPYDVPDYARDPAGWQDVYKRQXXX)

Fɪɢ. 2. Transposon ISΩ/hah. Cre recombinase excises the internal portion of the transposon, leaving an insertion of 162 nucleotides. For insertions that are in the same reading frame as the target gene, the small insertion comprises an open reading from of 54 codons. The nucleotide sequence and corresponding amino acid sequence are shown. Nucleotides in capital letters indicate the transposon end sequences. The loxP site is underlined. n, any nucleotide (produced by target-site duplication during transposition); strep/spec, streptomycin/spectinomycin; X, indeterminate residue encoded by target-site duplication.

(lacking antibiotic) previously spread with 10^6–10^7 phage $P1_{vir}$ and dried. Incubate overnight at $37°$. Infecting phage will express Cre recombinase, excising the portion of the transposon between the loxP sites.

4. Select colonies that exhibit increased blue color on the $P1_{vir}$ plate compared to the original plate (the color is usually punctate). Such colonies carry putative Cre-activated insertions at permissive sites in *lacZ*. Verify phenotypes by streaking for single colonies from the original streptomycin- or spectinomycin-resistant colony and patching isolated colonies on L agar supplemented with X-gal, IPTG, and $CaCl_2$, and previously spread with $P1_{vir}$.

5. LacZ$^+$ colonies may remain spectinomycin/streptomycin resistant due to unrecombined copies of the plasmid. Isolate plasmid from colonies showing phage activation and re-transform cells (e.g., CC118) to isolate cells carrying exclusively parental or recombined plasmid.

6. Identify insertion sites by DNA sequence analysis using primer CT1 (Table I) and assay β-galactosidase activities of parental and recombinant plasmids.

Large-Scale Transposon Mutant Library Construction

Large collections of unique insertion mutants can readily be generated using transposable elements, and it is possible to separately bank and sequence map the mutants to produce large, defined libraries with mutations in virtually every nonessential gene in a bacterial genome. Such libraries serve as valuable repositories of defined mutants for studying specific genes of interest, and can also be systematically screened to provide "complete" lists of genes responsible for a particular phenotype. They also help identify an organism's essential genes by exclusion (Judson and Mekalanos, 2000). Using transposons from Table I, we have assembled such libraries for *Pseudomonas aeruginosa* (Jacobs *et al.*, 2003), *Francisella novicida* (in preparation), *Escherichia coli*, and *Burkholderia thailandensis*. Figure 3A summarizes the process for constructing such libraries. The protocols below are derived primarily from our experience constructing the *P. aeruginosa* and *F. novicida* mutant libraries.

Generating and Arraying Mutants

If mutant libraries are created using transposons that generate reporter gene fusions, a choice must be made between arraying all insertion mutants or only the subset of mutants exhibiting reporter activity on the medium used for isolation. In general, we have included insertions regardless of their reporter activity in the interest of achieving comprehensiveness in

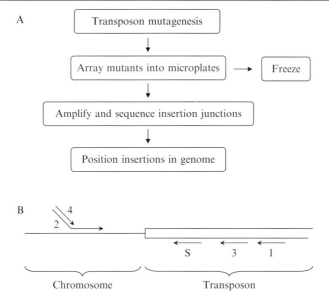

FIG. 3. Large-scale mutant library construction. (A) Schematic flow diagram for building and defining large-scale mutant libraries. (B) PCR amplification and sequencing of insertion junctions. In the first step, the transposon and adjacent chromosomal sequences are amplified using primers 1 (transposon specific) and 2 (a semidegenerate mixture with a unique sequence at the 5' end and four or five defined nucleotides at the 3' end). In the second step, the product of the first step is amplified using primers 3 (another transposon-specific primer) and 4 (corresponding to the 5' end of primer 2). In the third step, the product of the second step is sequenced using primer S (a transposon-specific primer). Refer to Table I for specific primers for each transposon.

the collection (e.g., including insertions in genes not expressed on the isolation medium). However, including the reporter-inactive insertions yields a collection in which the majority of such insertions are out-of-frame or in the incorrect orientation rather than being in-frame and fused to an unexpressed gene.

1. Mutagenize the bacterium of interest with a transposon of choice. This step must be tailored to the species being mutagenized. For example, mutagenesis of *P. aeruginosa* was carried out by introducing suicide plasmids carrying IS*lacZ*/hah or IS*phoA*/hah into cells pregrown at 42° rather than 37° to reduce restriction (Holloway, 1998). For *F. novicida*, a maximal yield of insertion mutants following electroporation of transposon–transposase complexes required lengthy outgrowth (>3 h) prior to plating on selective media. The most important general consideration arises from the need to minimize the number of sibling insertions generated. This is

done by reducing outgrowth to the minimum required to allow integration of the transposon and expression of the antibiotic-resistance marker, and by carrying out multiple independent mutageneses.

2. Plate the mutagenized culture onto nutrient agar supplemented with antibiotic to select for the transposon insertion, antibiotic to counter-select against the donor strain if conjugation is used, and an indicator if desired (X-Gal or XP). Plate an amount that will yield 100 to 200 colonies per 10-cm plate. Plating at a higher density produces many overlapping colonies, increasing the chances that individual colonies picked will not be pure. For robotic picking, the transformation mix is generally plated on large square plates (240 mm × 240 mm) such as Q-trays (Genetix). Incubate until colonies are large enough to be picked and arrayed.

3. Pick individual colonies into the wells of 96-well or 384-well plates filled with appropriate growth/freezer media supplemented with antibiotics as appropriate (Table II). For 384-well plates, we recommend square-well plates (e.g., Genetix X7007) filled with growth/freezer media (90 μl for Genetix X7007 plates). Colonies may be picked and arrayed manually or robotically (e.g., using a Genetix QPix2). For robotic picking, the parameters controlling robotic colony choice, pin sterilization, picking depth, well inoculation, and so on must be optimized for the individual species, colony sizes, and media and lighting conditions. Colonies that are colored due to expression of a reporter gene in the presence of a chromogenic indicator are difficult for robots that image in grayscale to recognize. In cases in which we wish to selectively array only colored colonies, we have patched the subset of colonies by hand onto separate plates prior to robotic arraying. In all cases, it is important to keep track of the original mutagenesis reactions from which the mutants arrayed into specific wells were derived.

4. Cover the 384-well plates with permeable tape cover (e.g., Qiagen Airpore Tape Sheet) and incubate overnight at 37°. To minimize evaporation

TABLE II
GROWTH/FREEZER MEDIA

Organism	Growth/freezer media[a]
B. thailandensis	LB freezing media[b], increase glycerol to 10%
E. coli	LB freezing media[b]
F. novicida	TSB freezing media[c] + 0.1% L-cysteine HCl + 0.2% dextrose
P. aeruginosa	LB + 5% DmSO

[a] Include antibiotics as used for mutant selection. For B. thailandensis and P. aeruginosa, antibiotics may interfere with strain longevity in frozen stocks (unpublished).
[b] Sambrook and Russell (2001)
[c] Recipe as for LB freezing media, but substitute tryptic soy broth for LB.

from the wells, keep the incubation chamber humidified by including open reservoirs of water. Optimal growth conditions vary depending on the organism. For *F. novicida*, shaking the plate during incubation improves growth, and an additional 24 h of incubation at room temperature without shaking is advised. For *P. aeruginosa*, shaking is not necessary, and too long of an incubation time often generates a thick pellicle, which makes future sampling difficult.

5. After the strains have grown to an acceptable turbidity, remove the permeable tape cover, cover the plates with their standard plastic lids and freeze at –80°. For long-term storage, it is also advisable to seal the plates using an aluminum tape seal (e.g., Island Scientific IS-200) or heat seal (IS-745). This prevents ice buildup on the top of the wells that may result in well-to-well contamination during thawing.

High-Throughput Mapping of Transposon Insertions

The transposon insertion sites of mutants arrayed in 96- or 384-well format can be efficiently identified using a high-throughput version of a semi-random, PCR-based method described earlier (Manoil, 2000). Figure 3B summarizes the amplification and sequencing method. An individual can routinely sequence map up to 3840 mutants per week with an average success rate of ~70%. The protocol presented below is designed for <KAN-2> insertions, but can be employed for other transposons by substituting the appropriate oligonucleotide primers (Table I). For all steps in the protocol, reactions and mixes should be kept on ice except when thermocycling. Move the plates from the ice into the thermocycler only when the thermocycler has reached the initial denaturation temperature. All centrifugations are at 4°.

1. Prepare PCR round 1 master mix (Table III), adding the reagents below in the order listed. Aliquot equally into 12 strip tubes, and use an electronic 12-channel pipettor to dispense 9.4 μl aliquots into all the wells of a 384-well PCR plate (e.g., MJ Research hard-shell plates). Using a 12-channel P-10 pipettor, dispense 0.6 μl from each well of a thawed freezer plate containing mutant cultures in freezer media into the wells of the PCR plate. This will require 384 sterile tips for the P-10 pipettor.

Seal the plate with MicroAmp adhesive film (Qiagen), briefly vortex to mix, centrifuge the plate for 1 min at 1000 rpm, and then carry out thermocycling as presented in Table IV.

2. Prepare PCR round 2 master mix (Table V) and aliquot 9.4 μl to each well of a 384-well PCR plate as described above. Transfer 0.6 μl of the PCR round 1 reactions to the round 2 PCR plate using either a 12-channel

TABLE III
PCR Round 1 Master Mix

Reagents for 10 μl PCR reaction	μl per reaction	μl for 96 reactions	μl for 384 reactions
H$_2$0	5.6	604.8	2419.2
10× TSG buffer	1	108	432
MgCl$_2$ 25 mM	1.5	162	648
dNTPs 10 mM	0.2	21.6	86.4
primer kan2-211, 10 μM	0.5	54	216
primer CEKG-2C 10 pmol/μl	0.17	18.36	73.44
primer CEKG-2D 10 pmol/μl	0.17	18.36	73.44
primer CEKG-2E 10 pmol/μl	0.17	18.36	73.44
TSG+ polymerase (Lamda Biotech)	0.1	10.8	43.2
Total	9.41	1016.28	4065.12

TABLE IV
PCR Round 1

Step	Temperature	Time	Notes
1	94°	12′	Colony denaturing
2	94°	30″	Denaturing
3	42°	30″	Annealing, decreased temperature 1° each cycle
4	72°	3′	Extension
5	Go to step 2		6 cycles, then to step 6
6	94°	30″	Denaturing
7	64°	30″	Annealing
8	72°	3′	Extension
9	Go to step 6		25 cycles, then step 10
10	72°	7′	Final extension
11	4°	Hold	

′, minutes; ″, seconds.

pipettor or a liquid-handling apparatus such as a Hydra II 384-pin liquid handler (Matrix). Liquid handler settings for accurate and consistent liquid transfer of low volumes should be optimized, and thorough pin washing between steps should be carried out according to the manufacturer's instructions. (Using the liquid handler for aliquoting the mutant cultures in step 1 is not advised, as the cells may clog or damage the dispensing pins.)

Immediately seal, vortex, and centrifuge the plate as described above, and then thermocycle as presented in Table VI.

3. Prepare cleanup reaction master mix (Table VII) and distribute 3 μl to each well of a 384-well PCR plate. Transfer 5 μl of each PCR round-2

TABLE V
PCR ROUND 2 MASTER MIX

Reagents for 20 μl PCR reaction	μl per reaction	μl for 96 reactions	μl for 384 reactions
H20	5.6	604.8	2419.2
10× TSG buffer	1	108	432
MgCl₂ 25 mM	1.5	162	648
dNTPs 10 mM	0.2	21.6	86.4
Primer kan2-145 10 μM	0.5	54	216
Primer CEKG-4 10 μM	0.5	54	216
TSG + polymerase	0.1	10.8	43.2
Total	9.4	1015.2	4060.8

TABLE VI
PCR ROUND 2

Step	Temperature	Time	Notes
1	94°	10′	Initial denaturing
2	94°	30″	Denaturing
3	64°	30″	Annealing
4	72°	3′	Extension
5	Go to step 2		30 cycles, then step 6
6	72°	7′	Final extension
7	4°	Hold	

′, minutes; ″, seconds.

TABLE VII
CLEANUP REACTION MASTER MIX

Reagents for 4 μl cleanup reaction	μl per reaction	μl for 96 reactions	μl for 384 reactions
Shrimp alkaline phosphatase (1U/μl; USB)	2	216	864
Exonuclease I (10 U/μl; USB)	1	108	432
Total	3.0	324	1296

reaction well to the cleanup reaction plate using a 12-channel pipettor or a liquid handler. Mix and centrifuge as described above, and then incubate in thermocycler as presented in Table VIII.

4. After the thermocycler incubations, add 20 μl of water to each well. Cleanup reactions may be stored frozen prior to sequencing.

5. Prepare sequencing master mix (Table IX) and distribute 6.4 μl to each well of the 384-well PCR plate. Transfer 1.6 μl of each cleanup reaction well to the sequencing reaction plate using a 12-channel pipettor or a liquid handler. Mix and spin as described above, and then thermocycle as shown in Table X.

6. Clean the reactions by precipitation prior to running on sequencer: Add 18 μl of 90% isopropanol to each well (for 8 μl sequencing reactions). Seal the plate, vortex to mix, then centrifuge at 3000 × g for 30 min. Unseal

TABLE VIII
CLEANUP REACTION

Step	Temperature	Time
1	37°	30′
2	80°	20′
3	4°	Hold

′, minutes.

TABLE IX
SEQUENCING MASTER MIX

Reagents for 8 μl sequencing reaction	μl per reaction	μl for 96 reactions	μl for 384 reactions
Water	3.4	367.2	1468.8
Dilution buffer	1	108	432
Primer kan2-125	1	108	432
BDT Version 3.1 (ABI)	1	108	432
Total	6.4	691.2	2764.8

TABLE X
SEQUENCING

Step	Temperature	Time	Notes
1	94°	5′	
2	94°	30″	
3	50°	10″	
4	60°	4′	
5	Go to step 2		30 cycles, then step 6
6	4°	Hold	

′, minutes; ″, seconds.

the plate, invert it onto a folded paper towel and centrifuge in the inverted position on the towel at 1000 rpm for 4 min. Add 20 μl of 70% isopropanol to each well. Do not vortex. Seal and centrifuge at 3000 $\times$ g for 15 min. Unseal the plate, invert onto a folded paper towel, and centrifuge inverted on the towel at 1000 rpm for 4 min. Incubate the plate open at 37° for 15 min.

7. Rehydrate the reactions by adding 18 μl of water to each well. Seal the plate using heat seal tape and a heat sealer (Eppendorf), vortex vigorously for 1 min, centrifuge for 1 min at 1000 rpm, and run the sequencing reactions on an automated sequencer (e.g., ABI 3730).

8. Using automated computational analysis, assign the transposon insertion location for each mutant based on the sequence traces. Successful sequence traces will show the junction between the transposon sequence and the flanking genomic DNA sequence. After computationally parsing out the transposon-specific sequence, the exact nucleotide of insertion can be determined by aligning (crossmatch) the genome-specific sequence to a genome sequence file. The specific gene interrupted can be determined by comparison to an annotation table. Because of the 9-bp duplication naturally created by Tn5 transposition, the effective nucleotide of insertion at the other end of the transposon will differ by nine base positions. Databases are maintained that catalogue all the sequence-mapped insertions in a given genome and their corresponding positions in the 96- or 384-well storage plates.

Acknowledgments

We thank M. Jacobs, T. Kawula, M. Olson, and I. Thaipisuttikul for numerous contributions. Research was supported by the National Institutes of Health Research Center for Excellence (grant 1-U54-A1-57141) and the Defense Advanced Research Projects Agency (contract N66001-02-1-8931).

References

Bailey, J., and Manoil, C. (2002). Genome-wide internal tagging of bacterial exported proteins. *Nat. Biotechnol.* **20,** 839–842.

Goryshin, I. Y., Jendrisak, J., Hoffman, L. M., Meis, R., and Reznikoff, W. S. (2000). Insertional transposon mutagenesis by electroporation of released Tn5 transposition complexes. *Nat. Biotechnol.* **18,** 97–100.

Hayes, F. (2003). Transposon-based strategies for microbial functional genomics and proteomics. *Annu. Rev. Genet.* **37,** 3–29.

Herrero, M., de Lorenzo, V., and Timmis, K. N. (1990). Transposon vectors containing non-antibiotic resistance selection markers for cloning and stable chromosomal insertion of foreign genes in gram-negative bacteria. *J. Bacteriol.* **172,** 6557–6567.

Holloway, B. W. (1998). 1998 Kathleen Barton-Wright Memorial Lecture. The less travelled road in microbial genetics. *Microbiology* **144**(Pt. 12), 3243–3248.

Jacobs, M. A., Alwood, A., Thaipisuttikul, I., Spencer, D., Haugen, E., Ernst, S., Will, O., Kaul, R., Raymond, C., Levy, R., Chun-Rong, L., Guenthner, D., Bovee, D., Olson, M. V., and Manoil, C. (2003). Comprehensive transposon mutant library of *Pseudomonas aeruginosa*. *Proc. Natl. Acad. Sci. USA* **100**, 14339–14344.

Judson, N., and Mekalanos, J. J. (2000). Transposon-based approaches to identify essential bacterial genes. *Trends Microbiol.* **8**, 521–526.

Maloy, S. R., Taylor, R. K., and Stewart, V. J. (1996). "Genetic Analysis of Pathogenic Bacteria: A Laboratory Manual." Cold Spring Harbor Laboratory Press, Cold Spring Harbor, NY.

Manoil, C. (1990). Analysis of protein localization by use of gene fusions with complementary properties. *J. Bacteriol.* **172**, 1035–1042.

Manoil, C. (2000). Tagging exported proteins using *Escherichia coli* alkaline phosphatase gene fusions. *Methods Enzymol.* **326**, 35–47.

Manoil, C., and Bailey, J. (1997). A simple screen for permissive sites in proteins: Analysis of *Escherichia coli* lac permease. *J. Mol. Biol.* **267**, 250–263.

Manoil, C., and Beckwith, J. (1985). TnphoA: A transposon probe for protein export signals. *Proc. Natl. Acad. Sci. USA* **82**, 8129–8133.

Manoil, C., Mekalanos, J. J., and Beckwith, J. (1990). Alkaline phosphatase fusions: Sensors of subcellular location. *J. Bacteriol.* **172**, 515–518.

Manoil, C., and Traxler, B. (2000). Insertion of in-frame sequence tags into proteins using transposons. *Methods* **20**, 55–61.

Mecsas, J. (2002). Use of signature-tagged mutagenesis in pathogenesis studies. *Curr. Opin. Microbiol.* **5**, 33–37.

Sambrook, J., and Russell, D. W. (2001). "Molecular Cloning: A Laboratory Manual," 3rd ed. Cold Spring Harbor Laboratory Press, Cold Spring Harbor, NY.

Shea, J. E., Santangelo, J. D., and Feldman, R. G. (2000). Signature-tagged mutagenesis in the identification of virulence genes in pathogens. *Curr. Opin. Microbiol.* **3**, 451–458.

Silhavy, T. J. (2000). Gene fusions. *J. Bacteriol.* **182**, 5935–5938.

Silhavy, T. J., and Beckwith, J. R. (1985). Uses of lac fusions for the study of biological problems. *Microbiol. Rev.* **49**, 398–418.

Zhou, M., Bhasin, A., and Reznikoff, W. S. (1998). Molecular genetic analysis of transposase-end DNA sequence recognition: Cooperativity of three adjacent base-pairs in specific interaction with a mutant Tn5 transposase. *J. Mol. Biol.* **276**, 913–925.

[13] Use of Operon and Gene Fusions to Study Gene Regulation in *Salmonella*

By KELLY T. HUGHES and STANLEY R. MALOY

Abstract

Coupling the expression of a gene with an easily assayable reporter gene provides a simple genetic trick for studying the regulation of gene expression. Two types of fusions between a gene and a reporter gene are possible. Operon fusions place the transcription of a reporter gene under the control of the promoter of a target gene, but the translation of the reporter gene and target gene are independent; gene fusions place the

METHODS IN ENZYMOLOGY, VOL. 421   0076-6879/07 $35.00
DOI: 10.1016/S0076-6879(06)21013-9

transcription and translation of a reporter gene under the control of a target gene, and result in a hybrid protein. Such fusions can be constructed *in vitro* using recombinant DNA techniques or *in vivo* using transposon derivatives. Many different transposon derivatives are available for constructing operon and gene fusions, but two extremely useful fusion vectors are (1) Mu derivatives that form operon and gene fusions to the *lacZ* gene, and (2) Tn5 derivative that forms gene fusions to the *phoA* gene.

Introduction

For decades *lac* operon and *lacZ* gene fusions have proved valuable genetic tools to study transcriptional and translational control of a wide variety of regulatory systems. Gene fusions to *phoA* and *lacZ* are important tools for the *in vivo* characterization of inner membrane protein topology, protein structure, and secretion. More recently the introduction of TEV protease sites has been included to study the topology of outer membrane proteins. This chapter focuses on the standard methods of *lac* and *phoA* fusions for the characterization of transcriptional and translational control and *phoA* in order to characterize inner membrane protein topology and secretion into the periplasm. This chapter is an addendum to the excellent descriptions of uses related to *in vivo* genetic engineering with derivatives of phage Mu (Groisman, 1991) and genetic fusions as experimental tools (Slauch and Silhavy, 1991) described in a previous volume of this series. The aim of this chapter is to focus on *Salmonella* genetic tools.

Mud Fusions

The Transposable Elements, MudJ and MudK

The transposons described here for constructing *lac* fusions in *Salmonella* are called MudJ and MudK, also known as MudI1734 and MudII1734, respectively (Groisman, 1991). These fusion vectors are derived from bacteriophage Mu (Symonds *et al.*, 1987). Like the bacteriophages lambda and P22, Mu can grow lytically on its host or be maintained as a lysogen. Bacteriophage lambda and P22 integrate by a site-specific recombination mechanism at specific attachment sites in the *Escherichia coli* or *Salmonella typhimurium* chromosome, respectively. Unlike lambda and P22, Mu is also a transposon that integrates by transposition at random sites in the host chromosome (Mizuuchi, 1992). When Mu inserts into a gene, it disrupts the gene and hence causes a mutation (the name Mu comes from mutator phage) (Taylor, 1963). Mu insertions are also polar on downstream genes in an operon. For example, if Mu inserts into the *lacZ* gene of *E. coli*, not only is the *lacZ* gene mutated, but the *lacY* and *lacA* genes are not expressed even though these genes are otherwise intact.

Mu derivatives, called Mud, have been constructed that carry the *lac* operon (without its promoter) near one end of Mu. These Mud transposons are lab-engineered elements in the sense that they consist of several pieces of DNA assembled using techniques of recombinant DNA. The original constructions were made by Malcolm Casadaban and colleagues (reviewed in Groisman, 1991).

In order to use the *lac* system to study transcriptional and translational regulation, two Mu*d* derivatives are generally used. One type (e.g., MudI, MudA, and MudJ) is used to make *lac* operon fusions to the promoter of a gene of interest. When this Mu derivative (Mud) is inserted in a gene in the correct orientation, the *lac* genes are expressed from the promoter of the mutated gene. Thus, expression of the *lac* operon is directly proportional to expression of the mutated gene. Since expression of the *lac* operon can be easily detected on indicator plates and quantified by assaying β-galactosidase activity, the expression of the mutant gene can be easily studied *in vivo*. A second type of Mud transposon (e.g., MudII, MudB, and MudK) is used to make *lacZ* gene fusions to a gene of interest. In this case, the Mu*d* insertion must be in the correct orientation within the inserted gene and in the correct reading frame to generate a Lac$^+$ fusion.

MudI and MudII are the original Mud-*lac* operon and Mud-*lac* gene fusion vectors constructed (Casadaban and Chou, 1984; Casadaban and Cohen, 1979). These vectors could transpose into genes resulting in the desired *lac* gene and operon fusions that facilitate the study of transcriptional and translational control of the gene of interest. However, because these fusion vectors retained the Mu transposase functions, they were not useful in selecting for regulatory mutants. Selection for regulatory mutants with higher-level expression of *lac* in a given gene of interest would select for transposition events that placed the Mud transposon under a more highly expressed promoter unrelated to the gene of interest. This was solved first by the isolation of MudA and MudB elements (Hughes and Roth, 1984, 1985). MudA and MudB are derivatives of MudI and MudII, respectively, in which the transposase functions are defective except in the presence of an amber suppressor mutation. When either MudA or MudB is introduced into a strain carrying an amber suppressor mutation, transposon occurs allowing for the isolation of *lac* fusions to genes of interest. Movement of the newly isolated fusions to strains lacking an amber suppressor allows for the selection of regulatory mutants. Later, mini-Mud vectors, MudJ (*lac* operon fusion) and MudK (*lac* gene fusion), were constructed in the Casadaban lab that were completely deleted for the transposition functions (Groisman, 1991).

(1) *lac* transcriptional fusion Mud vector

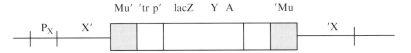

(2) *lac* translational fusion Mud vector
<u>Rationale</u>

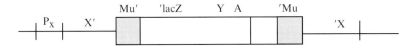

FIG. 1. Selecting for the appropriate Mud-encoded antibiotic resistance.

Each Mu*d* vector also possesses a gene that encodes resistance to an antibiotic. MudI, MudII, MudA, and MudB encode resistance to ampicillin while MudJ and MudK elements are resistant to kanamycin. Thus, transposition events are directly selected for by selecting for the appropriate Mud-encoded antibiotic resistance (Fig. 1).

In this experiment you will use a positive selection scheme to isolate *lac* fusions to genes required for biosynthesis of amino acids, nucleic acid bases, and selected vitamins in *Salmonella*. The transposable elements MudJ and MudK will be introduced into the *Salmonella* prototrophic strain LT2 by P22 transduction. Because the MudJ and MudK elements lack transposase, a special donor strain will be used to provide transposase both transiently and *in cis* to the MudJ and MudK elements. The MudJ donor strain carries both MudJ and MudI adjacent to each other in the *his* biosynthetic operon. The MudK donor strain carries both MudK and MudI adjacent to each other in the *his* operon. The insertions are oriented such that the transposase genes within the MudI element are proximal to either the MudJ or MudK element. MudI is 37 kilobase pairs in length; MudJ and MudK are 11 and 10 kbp, respectively, and each is located 5 kbp from the MudI element within the *his* operon.

P22 will package 44 kbp. This is not enough to package MudJ or MudK and MudI in the same transduced fragment. However, it is enough to package an intact MudJ or MudK, and the portion of MudI encoding transposase. When P22 is grown on the donor strains and used to transduce MudJ or MudK encoded kanamycin resistance, the transduced fragment enters the host cell, transposase is expressed *in cis*, and acts on the adjacent

Mud transposon ends. Transposase excises the Mud element and inserts it essentially at random into the chromosome. Then, the DNA fragment expressing transposase is degraded. The donor phage and recipient cell mixture is plated on rich medium containing kanamycin (Km). About 80% of the Km-resistant (Km^R) transductant colonies that form have inherited the MudJ or MudK elements by transposition.

The phage, grown on the MudJ and MudK donor strains, is mixed with recipient cells and allowed to stand for an hour on the bench and plated on rich medium containing kanamycin. The 1-h period allows time for recipient cells that have inherited MudJ or MudK elements by transposition to express the Km-resistance gene (phenotypic expression). After colonies have formed, they are replica printed to minimal medium (E + Km) and rich medium (Luria broth [LB] + Km). Direct comparison of these plates allows for the identification of auxotrophs (colonies that fail to grow on minimal medium).

If a cell expresses the kanamycin-resistance gene, Km^R, then it can grow despite the presence of kanamycin. Cells lacking such a gene are killed by kanamycin. In MudJ and MudK, the Km^R gene is expressed constitutively (continuously, even if *lac* is not expressed). No matter how or where the Mud inserts into the chromosome, the host cell acquires kanamycin resistance.

Use of Phage P22 to Transduce Mud

Phage P22 is a *Salmonella* phage that can perform generalized transduction. That is, when P22 multiplies in a cell, it occasionally packages fragments of the host chromosome within its particles. If MudJ or MudK is present in the host chromosome, these sequences can be packaged by P22 with flanking chromosomal sequences and transduced into the next recipient strain exposed to the phage stock. When Mud sequences are part of a transduced fragment, one of two things can happen. First, the Mud element can transpose or hop from the fragment to a randomly selected site in the recipient chromosome. (This action requires the presence of Mu transposase functions.) When this happens, a new insertion mutation is generated in the recipient strain. The second possibility is that the Mud sequences can remain in place and can be inherited by homologous recombination between the sequences flanking Mud in the fragment and the homologous sequences in the recipient chromosome. When this happens, no new mutation is generated, but the recipient inherits the same insertion mutation that was present in the donor strain on which P22 was grown.

We can control which of these events occur. In addition to ends of Mu, transposition requires several Mu proteins whose genes are not included in

MudJ or MudK. These proteins must be supplied if Mud is to transpose. In the experiment described below, the donor MudJ and MudK phage lysates are grown on strains containing both a MudI with functional transposase genes and a MudJ or MudK prophage closely linked to each other in the chromosome (Hughes and Roth, 1988). P22 fragments will frequently carry all of Mud and the transposition genes of the nearby Mu prophage. These Mu genes can be expressed and produce proteins that cause Mud to transpose from the transduced fragment to the recipient chromosome.

Thus, transposition is possible initially and all mutants isolated will inherit Mud by transposition. However, they will lose the piece of DNA that encodes the transposition functions and the Mud inserts will not be able to transpose again. In subsequent crosses, involving these insertion mutations, they can be inherited only by homologous recombination.

The lac *Operon and Use of X-gal MacConkey and TTC Indicators*

The *lacZ* gene encodes the structure of the enzyme β-galactosidase. Normally this enzyme catalyzes cleavage of the disaccharide lactose to its two constituent sugars glucose and galactose. This splitting is required if cells are to use lactose as a source of carbon and energy. In addition to this normal activity, this enzyme can catalyze splitting of galactose from a variety of organic molecules. One such compound has galactose joined to a dye molecule (abbreviated X). The compound X-gal is colorless but, when it is split by β-galactosidase (the *lacZ* gene product), it releases the free dye (X) which is deep blue.

This is extremely useful. If X-gal is incorporated into the growth medium in a petri dish, colonies growing on that medium turn blue *if and only if* they are expressing a *lacZ* gene. Because of its sensitivity, the use of X-gal as an indicator is most useful to distinguish the presence or absence of β-galactosidase (β-gal) activity. It can be used to distinguish different levels of β-gal activity at lower levels from 0 to about 30 β-gal units of activity. Minimal salts glucose medium provides a better background to visualize differences in β-gal levels than rich media such as LB. For differences between levels of β-gal activity up to about 200 β-gal units, MacConkey-lactose and tetrazolium-lactose indicator plates can be used. The dye triphenyl tetrazolium chloride (TTC) can be used to distinguish different levels of β-gal activity, either an indicator of lactose fermentation, or as an indicator of the cells' reductive capacity. Use of TTC as an indicator of lactose fermentation is based on lowering the pH in the media (Lederberg, 1948). On TTC-lactose indicator plates, cells that ferment lactose are white, while cells that do not ferment lactose are red. TTC can also be used to detect the cells' reductive capacity (Bochner and Savageau, 1977). In this case, lactose utilization increases the cells' reductive capacity, and they are able to reduce tetrazolium, whereby it turns red. Cells unable to utilize lactose utilize protease peptone in the medium resulting in a low reductive capacity, the inability to reduce TTC, and formation of white colonies. Thus, depending on the selections or screens to be used it might be useful to use an indicator that requires low levels of β-gal activity to be detected such as X-gal, or selections for higher β-gal activity to be detected as red (MacConkey-lactose or TTC-lactose reductive medium) or white (TTC-lactose fermentation medium).

The Plan of the Experiment

The Mutant Hunt for Auxotrophs

First, strain LT2 will be mutagenized with MudJ and MudK transposons selecting for Mud-encoded kanamycin resistance. This will be done by a transduction cross using P22 phage grown on a strain that carries MudJ or MudK sequences and adjacent Mu transposase genes (Hughes and Roth, 1988). Almost every Km^r colony that arises will be due to transposition of Mud from the transduced fragment into the recipient chromosome. Since each colony is the product of a different transposition event, each of the insertion mutants arise independently; no two should be alike. Also, the selections will be done on lactose indicator plates to screen for insertions that express β-galactosidase. For MudJ, these insertions will land in a gene in the correct orientation so that the promoter for that gene will transcribe the *lac* operon within MudJ. For MudK, these insertions will land in a gene in the correct orientation and the correct reading frame to create a *lacZ* fusion to the amino terminal portion of the protein encoded by the inserted gene. The insertion mutants will be screened for those that have inserted into auxotrophic genes by replica printing onto minimal glucose plates and LB plates. Km^R colonies that fail to grow on minimal plates will be kept as potential auxotrophic insertion mutants. The auxotrophic requirement will be determined by auxonography (described below).

Procedures

Procedure 1. Preparation of MudJ and MudK Donor
P22-Transducing Lysates

Strains

TT10288(*hisD9953*::MudJ*his-9944*::MudI)
TT10381(*hisD1284*::MudK*his-9944*::MudI)
LT2 (wild-type)

Media

LB: 10 g tryptone, 5 g yeast extract, and 5 g NaCl per liter of deionized water

LB agar: Same as LB with addition of 15 g agar per liter of deionized water

Ex50 salts (Maloy *et al.*, 1996)

50% D-glucose

Sterile saline: 8.5 g NaCl per liter deionized water

Top agar: 10 g tryptone and 7 g agar per liter deionized water

Motility agar: 10 g tryptone, 5 g NaCl, 3 g Bacto agar per liter deionized water

P22 broth: 200 ml LB, 2 ml Ex50 salts, 0.8 ml 50% D-glucose, 10^7 to 10^8 plaque-forming units (pfu)/ml of P22 transducing phage (P22 HT/*int*).

*P22 HT/*int *Lysate Preparations*

Grow 1 ml overnight cultures of strains TT10288 and TT10381 to saturation in LB. Add 4 ml of P22 broth to each culture and grow 5 to 36 h at 37° with shaking. Pellet cells by centrifugation (10 min full speed in a table top centrifuge, or for larger volumes spin 5 min at 8000 rpm in a SS34 rotor). Decant the supernatant into a sterile tube, add $CHCl_3$, and vortex to sterilize. Store at 4°.

Procedure 2. Isolation of Auxotrophic MudJ/K Insertion Mutants

Materials

P22 HT/*int* transducing lysates on *S. enterica* MudJ and MudK donor strains TT10288 and TT10381 (Hughes and Roth, 1988)

Recipient wild-type *S. enterica* strain LT2

1. Start a 1-ml overnight culture of the wild-type strain LT2. Grow overnight with aeration at 37°.

2. First, do a test cross to determine the number of MudJ or MudK insertions obtained per 0.1 ml of diluted phage stock. Dilute the MudJ and MudK donor lysates 10^{-2}, 10^{-3}, and 10^{-4}. In a sterile tube, mix 0.1 ml of cells from the overnight culture with 0.1 ml of phage grown on the MudJ or MudK donor. Some of the phage particles will inject Mud DNA and the adjacent Mu transposase genes from the linked MudI insertion into your recipient cells. This experiment will require replica printing to screen for insertions in auxotrophic genes; 300 to 500 colonies per plate is a good working number. The size of the target will determine the frequency at which a MudJ or MudK insertion is obtained. For a 1-kbp target gene, at least 5000 Km^R will need to be screened to obtain each MudJ insertion

mutant, and 10,000 for each MudJ inserted in the correct orientation to place the *lac* operon under control of the promoter of the gene into which MudJ has inserted. MudK insertions require both correct orientation and translational reading frame, or 30,000 Km^R insertion mutants will need to be screened to obtain a Lac^+ MudK insertion in a 1-kbp target gene.

3a. Start a fresh 20 ml overnight culture of strain LT2. Grow overnight with aeration at 37°. For MudJ insertion mutants, add 11 ml of cells to 11 ml of diluted phage using the dilution that gave 300 to 500 Km^R transductants per plate. Let sit 1 hr at room temperature to allow for phenotypic expression of Km^R. Plate 0.2 ml onto each of 100 LB plates with added Km (50 μg/ml). Incubate overnight at 37°.

3b. Since MudK insertions have to be in the correct orientation and reading frame, more colonies will have to be screened to obtain LacZ protein fusions (translational fusions) to genes of interest. Start a fresh 60-ml overnight culture of the *S. enterica* recipient strain. Add 55 ml of cells to 55 ml of diluted phage using the dilution that gave 300 to 500 Km^R transductants per plate. Let sit 1 h at room temperature to allow for phenotypic expression of Km^R. Plate 0.2 ml onto each of 500 LB plates with added Km (50 μg/ml). Incubate overnight at 37°.

4. For auxotrophic insertion mutants, replica-print LB-Km transduction plates to the minimal E glucose medium (Maloy *et al.*, 1996), Km (125 μg/ml), and LB Km (50 μg/ml) plates.

5. Read replica prints, and using a toothpick, pick the MudJ/K auxotrophic insertions and isolate P22-sensitive mutants using green indicator plates (Maloy *et al.*, 1996).

6. Using sterile toothpicks, pick and patch auxotrophic insertions onto LB plates. Replica print to the 11 diagnostic supplemented-minimal-media plates to determine the probable auxotrophic requirement.

7. Score plates for the auxotrophic requirements of individual mutants. Inoculate overnight cultures of individual mutants.

8. Pellet overnight cultures and resuspend in saline solution. Plate 0.1 ml onto a minimal-glucose-Xgal plate. Plate 0.1 ml onto a minimal-lactose plate. Towards one end of the plate, add a crystal of the potential auxotrophic requirement (based on the growth pattern on the 11 diagnostic supplemented minimal media plates) to each plate. Do this by breaking a sterile stick in half and use the sterile end to scoop up a crystal or small amount of powder for the auxotrophic supplement to be tested. If an individual auxotroph has multiple requirements, allow enough spacing between the different crystals added to the plate for growth to occur in between where enough of both supplements are present to allow for growth.

9. Score plates for auxotrophic requirement and regulation of gene expression. If the promoter is induced by starvation of the auxotrophic requirement, a dark blue halo will form as the cells starve for the supplement.

Nutritional Supplements

Table I contains a list of nutritional supplements used to identify specific auxotrophs. Stock solutions are such that 5 ml of the solution added to 1 liter of medium will result in the appropriate concentration (200×). Limiting concentration is the concentration at which auxotrophs requiring that supplement will form tiny colonies distinguishable from wild-type.

DIAGNOSIS OF AUXOTROPHS (AUXANOGRAPHY). The composition of these plates is described in Table II. All nutrients are used at the final concentrations. The compositions of media 1 to 5 are listed vertically in the table. The compositions of media 6 to 10 are listed horizontally. Medium 11 is an assortment of compounds not included in the others; its contents are listed horizontally at the bottom of the table. Some notes on the use of these media follow the table. Note that many compounds do not survive the autoclave. Thus, PABA, DHBA, PHBA, DAP, glutamine, and asparagine should be added to the LB medium after it is autoclaved.

Discussion

1. Some purine mutants grow on adenosine or guanosine; they will grow on pools 1, 2, and 6.
2. Some purine mutants require adenosine + thiamine; they will grow only on pool 6.
3. *pyrA* mutants require uracil + arginine; they grow on pool 9.
4. Mutants requiring isoleucine + valine will grow only on pool 7. When using *E. coli* K12, pool 5 must have isoleucine added, as all K12 strains are sensitive to valine in the absence of isoleucine.
5. Mutants with early blocks in the aromatic pathway will only grow on pool 8. In addition to the nutrients listed, pool 8 contains PABA, PHBA, and DHBA to satisfy mutants blocked in new synthesis of aromatic amino acids.
6. Early blocks in the lysine pathway also require DAP, and grow only on pool 4.
7. Pool 11 is a catchall, mostly vitamins. This list can be added to.
8. Solutions of the above nutrient pools (1 to 11) can be made up as a 10-fold concentrate over the final concentration used in media.
9. Use salts of glutamic and aspartic acids.
10. Do not autoclave glutamine or asparagine solutions.
11. Keep solutions containing tryptophan and tyrosine dark (wrapped in foil).

TABLE I

Nutritional Supplements

Nutrient	Plate conc (mM)	Low conc (mM)	Stock solution (%)	ml/l for low (%)	Sterilize	Remarks
Adenine	5.0	0.001	1.35		Filter (F)	0.1 N HCl
Adenosine	5.0	0.001	2.67		F	
Alanine	0.47		0.84		Autoclave (A)	
Arginine	0.6	0.01	2.53	0.86	F	
Asparagine	0.32		0.84		F	
Aspartate-K	0.3		1.0		F	
Biotin	0.1		0.49		F	
Cysteine	0.3		0.73		F	
Diaminopimelic acid (DAP)	0.1		0.38		F	
2,3-Dihydroxy-benzoic acid (DHBA)	0.04		0.06		F	
Glutamate-Na	5.0				F	
Glutamine	5.0		14.6		F	
Glycine	0.13		0.2		A	
Guanine	0.3		0.91		F	1 N HCl
Guanosine	0.3		1.7		F	
Histidine	0.1	0.005	0.31	0.25	A	
Histidinol	1.0		4.28		A	
Isoleucine	0.3		0.79		A	
Leucine	0.3	0.005	0.79	0.086	A	
Lysine	0.3	0.005	1.1	0.086	A	
Methionine	0.3		0.9	0.086	A	
Nicotinic acid	0.1		0.25		A	
Pantothenate-Ca	0.1		0.48		F	
p-Amino benzoic acid (PABA)	0.04		1.0		F	
p-Hydroxy benzoic acid (PHBA)	0.04		1.0		F	
Phenylaline	0.3		0.99		A	0.01 N HCl
Proline	2.0	0.002	4.6	0.005	A	
Pyridoxine-HCl	0.1		0.41		F	
Serine	4.0	0.01	8.4	0.0125	A	
Thiamine	0.05		0.337		A	
Threonine	0.3		0.71		A	
Thymine	0.32		0.81		F	
Tryptophan	0.1		0.41		F	
Tyrosine	0.1		0.36		F	
Uracil	0.1	0.003	0.224	0.15	A	
Uridine	0.1	0.003	0.488		F	
Valine	0.3		0.7		A	
EGTA	10.0		2M		A	Neutralize to pH7

TABLE II
AUXONAGRAPHY POOLS

	1	2	3	4	5
6	adenosine	guanosine	cysteine	methionine	thiamine
7	histidine	leucine	isoleucine	lysine	valine
8[a]	phenylalaine	tyrosine	tryptophan	threonine	proline-PABA, DHBA, PHBA
9[b]	glutamine	asparagine	uracil	aspartic acid	arginine
10	thymine	serine	glutamic acid	DAP	glycine
11	pyridoxine, nicotinic acid, biotin, pantothenate, alanine				

[a] See note 5 in "Discussion" section.
[b] Use 5 mM Gln in pool 1 and 20 mM Gln in pool 9 to isolate glnA mutants.

12. Pool 9 contains 20 mM of glutamine. Pool 1 contains 5 mM. Bacteria requiring high glutamine (such as *glnA* mutants) will grow only on pool 9.

Procedure 3. Isolation of MudJ/K Insertions in Flagellar Genes

Introduction: The Flagellum

Flagella are reversible rotary devices that act as propellers allowing bacteria to "swim" through liquid environments and crawl along surfaces. When coupled to the sensory pathway involved in measuring changing nutrient concentrations in the media, bacteria are able to migrate toward a higher concentration of carbon and energy sources or away from harmful chemicals. This behavior is referred to as chemotaxis. If the flagella are rotated in a counterclockwise direction, the individual flagella form a bundle and the cell is propelled forward (smooth swimming). If the direction of flagellar rotation switches to a clockwise direction, the bundle comes apart and the cell tumbles. Upon switching back to smooth swimming, the cell is headed off in a different direction. By controlling the amount of time spent smooth swimming and tumbling (switching bias), the cell reaches its destination in a biased, random-walk fashion. Aside from their role in bacterial behavior, flagella are used as models for other cellular processes including the regulation of a large gene system, assembly of a complex bacterial organelle-like structure, and the coupling of gene expression to the assembly of this structure. An enormous advantage to other organelle systems is the fact that bacterial motility is completely dispensable to the viability of the cell in the lab, and is thus amenable to genetic dissection.

Over 60 genes have been identified in the motility process. These include the Fla class (*flg, flh, fli,* and *flj* genes; whether a *fla* gene is designated *flg, flh, fli,* or *flj* depends on its location in the bacterial chromosome), which code for proteins required for the assembly of the flagella; the Che class, which are involved in the chemotactic response (environmental sensing and the switching bias in direction of flagellar rotation); and the Mot class, which are the motor force generators. The flagellar motor is driven by a proton motive force. The following procedure describes the isolation of MudJ/K insertions in just the Fla class of genes using a positive selection method.

The Flagellar Regulatory Hierarchy

The regulation of flagellar gene expression is coupled to the assembly of the flagellar organelle. In *S. typhimurium*, the flagellar regulon comprises over 50 genes and is organized into a transcriptional hierarchy of three promoter classes (reviewed in Chilcott and Hughes, 2000). The class 1 promoter transcribes a single operon, the *flhDC* operon. The FlhC and FlhD proteins form a heterotetrameric complex, $FlhC_2FlhD_2$, that activates σ^{70}-dependent transcription from class 2 promoters. The products of class 2 transcripts are primarily required for the structure and assembly of the hook-basal body (HBB) structure. Among class 2 transcribed genes is the *fliA* gene that encodes the flagellar-specific transcription factor σ^{28}. The σ^{28} holoenzyme of RNA polymerase transcribes class 3 promoters. In general, class 3 transcripts code for proteins required late in the flagellar assembly process and genes that code for the chemosensory system.

Negative regulatory proteins, FlgM and FliT, coordinate the transition from HBB completion to initiation of class 3 transcription. FlgM is an anti-σ^{28} factor that inhibits σ^{28}-dependent transcription from class 3 promoters prior to HBB completion. FliT is a negative regulator that inhibits class 2 transcription by direct interaction with the FlhDC activation proteins (Yamamoto and Kutsukake, 2006). Prior to HBB completion, FliT binds to the cap protein FliD and is inactive as a negative regulator. Upon HBB completion, FlgM and FliD are secreted from the cell, and σ^{28} is free to transcribe the late assembly genes now needed, and FliT is free to further HBB, class 2 gene transcription. In this way, genes whose products are assembled outside the cytoplasm in the final assembly stage (in particular, the large external filament) are not transcribed until the earlier assembly stage onto which these late subunits will be added is completed. Also, once the HBB is complete and genes encoding HBB components are no longer needed, they are turned off by FliT.

Isolation of MudJ/K Insertions in Genes Required for Flagella Assembly

Two positive selections will simultaneously be employed to isolate Mud insertions in the flagellar genes. MudJ/K will be introduced into a wild-type *S. typhimurium* recipient by P22-mediated transduction and select for MudJ-encoded kanamycin resistance. The second selection is to also have bacteriophage Chi present on the kanamycin-selection plates. Chi is a virulent phage (it only grows via a lytic cycle) that attaches to bacteria at the flagella, migrates to the base of the flagellar structure, and injects its DNA. If cells are defective in the flagella structure, then Chi cannot infect and the cells are resistant to killing by phage Chi. By doing a MudJ transposition experiment in the presence of bacteriophage Chi, only insertions that disrupt the flagellar genes will survive selection.

The Plan of the Experiment: The Mutant Hunt

First, *S. enterica* strain LT2 will be mutagenized with the MudJ or MudK transposons, selecting simultaneously for resistance to killing by phage Chi and for MudJ-encoded kanamycin resistance (Km^R). This will be done by a transduction cross using P22 phage grown on a strain that carries MudJ sequences. This phage will be used to transduce the wild-type recipient strain LT2. Selection is made for inheritance of the kanamycin-resistance gene of MudJ and loss of a flagellar gene by plating the mixture of transducing phage and recipient cells on plates containing both kanamycin and the virulent phage Chi. Almost every Km^R colony that arises will be due to transposition of MudJ from the transduced fragment into the recipient chromosome. Since each colony is the product of a different transposition event, each of the insertion mutants arose independently; no two should be alike. Also, the selections will be done on MacConkey-lactose indicator plates to screen for those insertions that have fused the *lac* operon to a promoter.

Resistance to phage Chi can occur by one of two means. One possibility is for the MudJ element to insert into a flagellar gene, which would result in resistance to infection by Chi. The second possibility is that a spontaneous flagellar mutation occurred, resulting in Chi resistance that is unrelated to the location of the Mud insertion.

Once putative *fla*::Mud insertions are isolated, they will be made phage-free on green indicator plates. They will then be checked for loss of flagellar function on a motility indicator plate. On such a plate, wild-type bacteria with functional flagella will swim outward from the point of inoculation producing a diffuse swarm colony (Mot^+); a mutant defective in flagella formation will produce a tight nonmotile colony on motility plates (Mot^-). Once it is established that the insertion mutants are defective in motility, a process called flair complementation will be used to determine which of three chromosomal locations of flagellar genes the Mud insertion has located into—*flg*, *flh*, or *fli* regions. Flair complementation (described below) is a time-honored process that uses abortive transduction to make individual cells diploid for the gene of interest. It was also the original demonstrations of complementation in bacteria.

Phage-transducing lysates grown on individual strains deleted for different flagellar genes. When these lysates are mixed with a culture of a *fla*::Mud insertion mutant, they can transduce the mutant to Mot$^+$ if and only if they are not deleted for the flagellar gene that the Mud has inserted into. When P22 introduces a DNA fragment from one strain into another, some fragments are recombined with the recipient chromosome. Many other DNA fragments are not recombined, and exist as a protein-bound circular DNA fragment that is not replicated (an abortive fragment). Upon cell division, only one of the two daughter cells receives the abortive fragment. If this fragment contains the functional flagellar gene that was lost by insertion of the Mud, then it will complement the defective gene and that cell will swim away from the site of inoculation toward new food sources. But each time the cell divides, only one of the two daughters will swim; the other gets left behind. This results in a trail of Mot$^-$ daughter cells from the original site of inoculation. This allows for the identification of which known flagellar deletion mutant will not complement an individual insertion mutant, and therefore which flagellar region the Mud has disrupted in the strain.

Materials

P22 HT/*int* transducing lysates on *S. enterica* MudJ and MudK donor strains TT10288 and TT10381

P22 HT/*int* transducing lysates on Fla$^-$ deletion strains for complementation analysis: SJW191 (*ΔflgA-J1191*), SJW1518 (*ΔflgG-L2157*), SJW1399 (*Δtar-flhD2039*), SJW1368 (*ΔflhA-cheA2018*), SJW1411 (*ΔfliA-D2050*), SJW1572 (*ΔfliE-K2211*), and SJW1556 (*ΔfliJ-R2195*).

P22 H5 virulent strain

Recipient wild-type *S. enterica* strain LT2

Preparation

1. Start an overnight culture of wild-type *Salmonella enterica* strain LT2 in 2 ml of LB. Grow overnight with aeration at 37°.

2. In a sterile tube, mix 0.5 ml of cells from the LT2 overnight culture with 0.5 ml of the MudJ or MudK donor phage lysate. Let stand at room temperature for 30 min. This allows phage infection. Add 4.5 ml of tryptone broth +Tc to the tube and incubate at 37° for an additional 30 min. This allows for expression of the Mud-encoded kanamycin-resistance gene and loss of expression of the gene into which the Mud has inserted. Transfer 1 ml of the phage and cell mixture from the tube into a separate sterile tube. Add Chi phage at a moi (multiplicity of infection) of 5 to the tube. Add the Chi phage + cells to 2.5 ml of molten top agar

(cooled to ~50°). Mix with brief vortex (at 5 setting) and quickly pour onto a MacConkey-lactose (Mac-Lac) + kanamycin plate. It is important to work quickly. *Do not let the agar stand at room temperature or it will solidify in the tube!* Let top agar solidify on the plate and incubate at 30°. Also, prepare a no-Chi control, a no-P22 control, and a no-cell control.

3. Once Km^R colonies arise, pick Km^R Chi^R Lac$^+$ colonies from the selection plates and streak for single colonies onto green indicator plates (Maloy *et al.*, 1996). Incubate at 37°.

4. Cross streak-potential phage-free clones against P22 H5 on green indicator plates. Incubate at 37°.

5. Pick phage-free clones off the green indicator plates (stay as far away from the H5 phage as possible), and streak for single colonies on LB + X-gal plates. Incubate at 37°.

6. Touch a single colony from each potential *fla* mutant with a toothpick and stab cells into a motility plate. Stab up to 24 potential Fla::MudJ mutants per plate. Incubate at 37°.

7. Any Fla$^+$ strains on your motility plates will eventually overgrow the plate. However, the Fla$^-$ mutants can be distinguished because they form dense growth where they were originally inoculated. If there is overgrowth on the motility plate from a false-positive isolate, simply retest the mutants on new motility plates, but leave off any that gave a Mot$^+$ phenotype. Inoculate a 2-ml LB culture for each bona fide *fla*::Mud insert to be tested by complementation assays. Incubate overnight at 37° with aeration.

8. Add 0.1 ml of each overnight culture to 0.1 ml of P22 phage grown on each of seven different *fla* deletion mutants. Let sit for 20 min at room temperature. Dip a sterile wooden applicator into each tube and apply a drop of the phage and cell mixture to the surface of a motility-Km plate. Try to tip the tube and swirl the stick around to pick up as much liquid as possible. For each Mud insertion mutant, place individual drops from four of the P22 infection tubes onto one plate and three on a second plate. Thus, there are two plates for each insertion mutant (eight plates total). Incubate at 37°. *Be gentle; do not gouge the surface of the agar.* Also, streak each *fla*::Mud insertion mutant onto L-Xgal, MacLac, and TTC-Lac plates to estimate the level of β-gal for each mutant.

9. Score plates for Lac expression and the appearance of trail complementation as described in the text. Lack of complementation indicates the location of the Mud insertion on the chromosome. If desired, the strain complementation may be repeated using phage lysates grown on nonpolar insertions in individual flagellar genes within a noncomplementing deletion interval to determine the specific gene that the Mud is inserted into.

Procedure 4. phoA *Gene Fusions*

Introduction

Alkaline phosphatase is encoded by the *phoA* gene in *E. coli*. The wild-type *phoA* gene has a signal sequence allowing export of alkaline phosphatase into the periplasm where it is active. Due to the highly reducing environment, alkaline phosphatase is not active in the cytoplasm. Alkaline phosphatase activity can be detected on solid media containing X-P (a colorless compound cleaved to form a blue colored compound like X-gal).

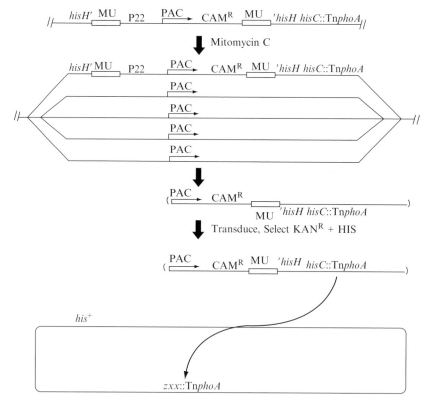

Fig. 2. Packaging of Tn*5-phoA* by MudP22. MudP22 cannot excise when induced by mitomycin C. Upon induction of the locked-in MudP22 prophage, *in situ* DNA replication of the MudP22 lysogen results in a localized amplification of the region of the chromosome where the MudP22 is inserted. Headful packaging initiates from the *pac* site within MudP22 and will include adjacent chromosomal DNA. In the case of the strain with a *hisH*::MudP22 and *hisC*::Tn*5-phoA*, induction of the *hisH*::MudP22 results in the packaging of the adjacent *hisC*::Tn*5-phoA*-containing DNA resulting in a high-frequency transducing lysate of the Tn*5-phoA* into a recipient cell.

Gene fusions to *phoA* provide an assay for extracytoplasmic proteins or domains of proteins (Manoil, 1990). If the PhoA protein missing its N-terminal secretion signal ('PhoA) is fused to a domain of a protein in the periplasm, it will result in alkaline phosphatase activity (yielding an X-P$^+$ colony). If 'PhoA is fused to a domain of a protein that is in the cytoplasm, it will *not* result in alkaline phosphatase activity, thus yielding an X-P$^-$ colony (Calamia and Manoil, 1990).

Use of MudP22 for Delivery of Transposon TnphoA

Tn*phoA* is a derivative of transposon Tn5 that generates *phoA* gene fusions (Manoil *et al.*, 1990). Tn5 transposes at a high frequency when it enters a naive host. Therefore, a strain with a TnphoA insertion immediately downstream of a MudP22 *pac* site was constructed for delivery of Tn*phoA*. A cartoon of this process is shown in Fig. 2.

Materials

Mud-P22 lysates on *S. enterica* Tn5 donor strain MS2001 (*hisG10085*:: Tn5-*phoA his-9956*::MudP22/ pJS28 (ApR P22 gene 9$^+$, or tail gene) (Wilson and Maloy, 1987)
Recipient wild-type *S. enterica* strain LT2

Preparation of MudP22 Lysates

1. Grow MudP22 strain overnight in LB broth + Ap (100 μg/ml).
2. Subculture 0.1 ml into 5 ml LB broth. Grow at 37° for 90 min with shaking.
3. Add 5 μl of 2 mg/ml stock of mitomycin C (Sigma). *Caution: Mitomycin C is a mutagen. Wear gloves and avoid contact.*
4. Incubate on 37° shaker for about 3 h or until the culture clears.
5. Pellet the debris by centrifuging for 1 min in microfuge tubes.
6. Decant supernatant into a screw capped test tube.
7. Add about 0.3 ml of CHCl$_3$ and vortex thoroughly. Store at 4°.

Transposition of TnphoA

1. Transduce a Kan^S recipient as described for P22 transduction.
2. Allow phenotypic expression, then plate on LB + Km + X-P.
3. Incubate the plates overnight at 37°.
4. X-P$^+$ colonies have insertions in secreted proteins.

References

Bochner, B. R., and Savageau, M. A. (1977). Generalized indicator plate for genetic, metabolic, and taxonomic studies with microorganisms. *Appl. Environ. Microbiol.* **33,** 434–444.

Calamia, J., and Manoil, C. (1990). lac permease of *Escherichia coli*: Topology and sequence elements promoting membrane insertion. *Proc. Natl. Acad. Sci. USA* **87,** 4937–4941.

Casadaban, M. J., and Chou, J. (1984). *In vivo* formation of gene fusions encoding hybrid beta-galactosidase proteins in one step with a transposable Mu-*lac* transducing phage. *Proc. Natl. Acad. Sci. USA* **81,** 535–539.

Casadaban, M. J., and Cohen, S. N. (1979). Lactose genes fused to exogenous promoters in one step using a Mu-*lac* bacteriophage: *In vivo* probe for transcriptional control sequences. *Proc. Natl. Acad. Sci. USA* **76,** 4530–4533.

Chilcott, G. S., and Hughes, K. T. (2000). The coupling of flagellar gene expression to flagellar assembly in *Salmonella typhimurium* and *Escherichia coli*. *Microbiol. Molec. Biol. Rev.* **64,** 694–708.

Groisman, E. A. (1991). *In vivo* genetic engineering with bacteriophage Mu. *Methods Enzymol.* **204,** 180–212.

Hughes, K. T., and Roth, J. R. (1984). Conditionally transposition-defective derivative of Mu*dl* (Amp Lac). *J. Bacteriol.* **159,** 130–137.

Hughes, K. T., and Roth, J. R. (1985). Directed formation of deletions and duplications using Mud(Ap, *lac*). *Genetics* **109,** 263–282.

Hughes, K. T., and Roth, J. R. (1988). Transitory cis complementation: A method for providing transposition functions to defective transposons. *Genetics* **119,** 9–12.

Lederberg, J. (1948). Detection of fermentative variants with tetrazolium. *J. Bacteriol.* **56,** 695.

Maloy, S. R., Stewart, V. J., and Taylor, R. K. (1996). "Genetic Analysis of Pathogenic Bacteria." Cold Spring Harbor Laboratory Press, Cold Spring Harbor, NY.

Manoil, C. (1990). Analysis of protein localization by use of gene fusions with complementary properties. *J. Bacteriol.* **172,** 1035–1042.

Manoil, C., Mekalanos, J. J., and Beckwith, J. (1990). Alkaline phosphatase fusions: Sensors of subcellular location (minireview). *J. Bacteriol.* **172,** 515–518.

Mizuuchi, K. (1992). Transpositional recombination: Mechanistic insights from studies of mu and other elements. *Annu. Rev. Biochem.* **61,** 1011–1051.

Slauch, J. M., and Silhavy, T. J. (1991). Genetic fusions as experimental tools. *Methods Enzymol.* **204,** 213–248.

Symonds, N., Toussaint, A., van de Putte, P., and Howe, M. (1987). "Phage Mu." Cold Spring Harbor Laboratory Press, Cold Spring Harbor, NY.

Taylor, A. L. (1963). Bacteriophage-induced mutation in *Escherichia coli*. *Proc. Natl. Acad. Sci. USA* **50,** 1043–1051.

Wilson, R. B., and Maloy, S. R. (1987). Isolation and characterization of *Salmonella typhimurium* glyoxylate shunt mutants. *J. Bacteriol.* **169,** 3029–3034.

Yamamoto, S., and Kutsukake, K. (2006). FliT acts as an anti-FlhD2C2 factor in the transcriptional control of the flagellar regulon in *Salmonella enterica* serovar typhimurium. *J. Bacteriol.* **188,** 6703–6708.

[14] Genomic Screening for Regulatory Genes Using the T-POP Transposon

By Changhan Lee, Christopher Wozniak, Joyce E. Karlinsey, and Kelly T. Hughes

Abstract

The identification of a gene that activates or regulates a gene or regulon of interest often requires the artificial induction of the regulatory gene. The properties of the Tn*10*-derived transposon T-POP allow a simple chromosomal survey of genes that, when artificially induced from an adjacent T-POP transposon by the addition of tetracycline, can activate or inhibit the expression of virtually any gene of interest. Procedures for genome-wide screening for T-POP inducible regulatory genes are described in detail. T-POP is a derivative of transposon Tn*10d*Tc. It encodes resistance to tetracycline, but unlike Tn*10d*Tc, the *tet*^{A} and *tet*^{R} promoters do not terminate within the transposon. Instead they continue out into adjacent chromosomal DNA. When this element inserts in a gene, three things will result: (1) the target gene is disrupted by the addition of a large block of DNA (approximately 3000 bases); (2) a drug-resistance gene (tetracycline resistance) included in the inserted material is now 100% linked to the insertion mutant phenotype; and (3) the mRNA transcripts initiated at either the *tet*^{A} or *tet*^{R} promoters (or both) will continue out into the adjacent chromosomal DNA. Despite the fancy aspects, insertion mutants are easy to isolate and can be assayed for effects on gene regulation using simple plate tests.

Introduction

The Transposable Element, T-POP

This chapter is an addendum to the excellent description of transposon uses, emphasizing Tn*10* and its derivatives, described in a previous volume of this series (Kleckner *et al.*, 1991). One of the early drawbacks associated with transposon Tn*10* was the tendency to transpose into particular sites within a gene, called hot spots. This was overcome through the isolation of Tn*10* transposase mutants with altered target specificity (ATS). The Tn*10*-ATS transposase has altered target specificity and inserts more randomly into a particular gene. The transposon described here is called T-POP. Most of the material is derived from transposon Tn*10d*Tc. Tn*10d*Tc is a trimmed-down version of Tn*10*. It lacks most of the IS*10* elements including the

METHODS IN ENZYMOLOGY, VOL. 421
0076-6879/07 $35.00
DOI: 10.1016/S0076-6879(06)21014-0

transposase gene. Thus, transposase must be expressed in the cell that Tn*10*-*d*Tc is introduced into so that the Tn*10d*Tc transposon can "hop."

The tetracycline-resistance gene of transposons Tn*10* and Tn*10d*Tc, *tet*^{A}, encodes a membrane protein that pumps tetracycline out of the cell, thereby conferring resistance to the antibiotic tetracycline. If a cell has this gene, then it can grow despite the presence of tetracycline. Cells lacking such a gene will not grow in the presence of tetracycline. The Tet^R protein represses *tet*^{A} and its own structural gene, *tet*^{R}. Tet^R binds to a region between the divergently transcribed *tet*^{A} and *tet*^{R} genes to repress transcription. When tetracycline is present in the media it binds the Tet^R protein and it becomes inactive. This results in transcription of both the *tet*^{A} and *tet*^{R} genes. The Tet^A efflux pump is incorporated into the cytoplasmic membrane and confers resistance to tetracycline.

The T-POP elements are derivatives of Tn*10d*Tc, constructed to remove transcription terminators at the ends of the *tet*^{A} or *tet*^{R} genes (Rappleye and Roth, 1997) (Fig. 1). In one version, the terminator at the end of the *tet*^{A} gene is removed so that those transcripts initiating from the *tetA* promoter continue into adjacent chromosomal DNA until a termination signal occurs. In a second version, the terminator at the end of the *tet*^{R} gene is removed so that those transcripts initiating from the *tet*^{R} promoter

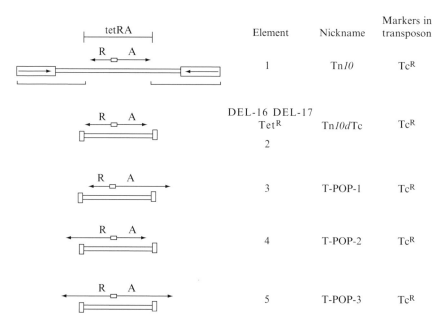

FIG. 1. Tn*10*, Tn*10d*Tc, and T-POP derivatives.

continue into adjacent chromosomal DNA until a termination signal occurs. In the third version, the terminators at the ends of both the tet^A gene and the tet^R gene are removed so that those transcripts initiating from both the promoters continue into adjacent chromosomal DNA until a termination signal occurs.

Use of Phage P22 to Transduce T-POP

The T-POP transposon was originally constructed for use in *Salmonella enterica*. Thus, bacteriophage P22 is the vector of choice for introduction of T-POP derivatives into cells. However, it would be relatively straightforward to introduce T-POP into broad host plasmids for transmission into other bacterial species. Care must be taken because introduction of the tet^A gene into high-copy-number plasmids is lethal when *tetA* is induced by addition of tetracycline. In this case, the tetracycline analog, oxytetracycline (Oxy-Tc), must be used as the selected antibiotic. This is because the basal level of Tet^A produced from tet^A cloned on high-copy-number plasmids is enough to confer resistance to Oxy-Tc, and Oxy-Tc does not inhibit Tet^R repression.

Phage P22 is a *Salmonella* phage that can perform generalized transduction. That is, when P22 multiplies in a cell, it occasionally packages fragments of the host chromosome within its particles. If T-POP is present in the host chromosome, these sequences can be packaged by P22 with flanking chromosomal sequences and be transduced into the next recipient strain exposed to the phage stock. When T-POP sequences are part of a transduced fragment, one of two things can happen. First, the T-POP element can transpose or hop from the fragment to a randomly selected site in the recipient chromosome if and only if Tn*10* transposase is expressed in the cell. When this happens, a new insertion mutation is generated in the recipient strain. The second possibility is that if no Tn*10* transposase is present, the T-POP sequences can remain in place and will be inherited by homologous recombination between the sequences flanking T-POP in the fragment and the homologous sequences in the recipient chromosome. When this happens, no new mutation is generated, but the recipient inherits the same insertion mutation that was present in the donor strain on which P22 was grown.

We can control which of these events occur. In addition to ends of Tn*10*, transposition requires Tn*10* transposase, whose gene is not included in T-POP. This protein must be supplied if T-POP is to transpose. Two methods are routinely used to introduce T-POP by P22 transduction. In the first method, the donor phage is grown on a strain containing T-POP inserted into an extrachromosomal DNA plasmid called the F factor. P22 fragments will frequently carry all of T-POP and the adjacent F plasmid DNA. If the recipient does not carry the F plasmid, T-POP cannot be inherited by homologous recombination and no tetracycline-resistant transductants will occur. If, however, the recipient carries another plasmid that expresses Tn*10* transposase, the T-POP element can transpose randomly into the recipient chromosome. The second method utilizes a P22 locked-in prophage, called Mu*d*P22, which is located adjacent to a T-POP element on the F plasmid. When this locked-in P22 prophage is induced, it replicates *in situ* resulting in a localized amplification of the region of the F plasmid, including the Mu*d*P22 element and adjacent chromosomal DNA. The Mu*d*P22 element is designed to package only DNA adjacent to the Mu*d*P22 element. When T-POP is inserted adjacent to Mu*d*P22, then the majority of P22 transducing particles produced following induction of Mu*d*P22 carry the T-POP element. This results in a 500-fold enrichment for transducing particles that carry T-POP.

Transposition is possible because the recipient strain that is used expresses transposase and all tetracycline-resistant mutants isolated will inherit T-POP by transposition. In subsequent crosses, involving cells that do not express transposase, they will be inherited by homologous recombination.

Use of T-POP to Identify Genes Affecting Regulation of Gene Expression

This method employs both a positive selection followed by a screening technique to isolate T-POP insertions that regulate the expression of a gene of interest. For the purpose of discussion, we will describe the isolation of T-POP insertions that regulate the *fliC* promoter fused to the *lac* operon using the Mu*d*J *lac* operon fusion vector. The *fliC* gene is the structural gene for the FliC flagellar filament protein.

The Mu*d*J element is a lab-engineered element in the sense that it consists of several pieces of DNA assembled using techniques of recombinant DNA. It contains a kanamycin-resistance operon and the *lac* operon. Mu*d*J makes *lac* operon fusions. When Mu*d*J is inserted in a gene in the correct orientation, the *lac* genes are expressed from the promoter of the mutated gene. For the Mu*d*J insertion in *fliC* gene described here, the *lac* operon is transcribed from the *fliC* promoter, and is subject to regulatory signals that affect flagellar gene transcription. Thus, expression of the *lac* operon is directly proportional to expression of the *fliC* gene. Since expression of the *lacZ* gene can be easily detected on indicator plates and quantified by assaying β-galactosidase activity, the expression of the *fliC* gene can be easily studied *in vivo*.

The T-POP transposon will be into a strain expressing Tn*10* transposase and carrying the *fliC*::Mu*d*J insertion by P22-mediated transduction.

Strains that inherit the T-POP element will be selected for by selecting for T-POP–encoded tetracycline resistance. These will then be screened for T-POP insertions that affect the expression of the *fliC-lac* fusion by looking for differences in *lac* expression (and thus *fliC* promoter transcription) in the presence and absence of tetracycline. Strains that have a T-POP insertion that are Lac⁻ in the presence of tetracycline and Lac⁺ in the absence of tetracycline carry T-POP insertions where the tet^A or tet^R promoter in T-POP is expressing a negative regulator of *fliC* gene transcription. Strains that have a T-POP insertion that are Lac⁺ in the presence of tetracycline and Lac⁻ in the absence of tetracycline carry T-POP insertions where the tet^A or tet^R promoters in T-POP are expressing a positive regulator of *fliC* gene transcription. This is then verified by growing P22 transducing phage on the *fliC*::MudJ strain carrying the T-POP insert and then using this lysate to transduce the T-POP into wild-type strain (LT2) that does not express Tn*10* transposase and does not harbor the *fliC*::MudJ insert. This will inherit the T-POP insert by homologous recombination. We can then test for the effect of the T-POP insertion on flagellar gene expression by screening for motility of the bacteria in the presence and absence of tetracycline. T-POP inserts that turned off the *fliC*::MudJ fusion in the presence of tetracycline should be nonmotile in the presence of tetracycline. T-POP inserts that were *fliC-lac* "off" in the absence of tetracycline, but *fliC-lac* "on" in the presence of tetracycline, will be motile only in the presence of tetracycline.

The Plan of the Experiment: The Mutant Hunt

First, T-POP is introduced into strain LT2 carrying a *fliC*::MudJ insertion, that also carries a pBR plasmid expressing transposase (pNK), with T-POP transposon selecting for resistance to T-POP–encoded tetracycline resistance. This is done by a transduction cross using P22 phage grown on a strain that carries T-POP sequences. This phage is used to transduce T-POP into the *fliC*::MudJ recipient strain. Cells are grown overnight to a final density of $\sim 2 \times 10^9$ ml. A 0.1-ml portion of an overnight culture is mixed with P22 phage grown on the T-POP donor strain. The amount of phage used will be designed to give 200 to 300 tetracycline-resistant colonies per plate. Selection is made for inheritance of the tetracycline-resistance gene of T-POP by plating the mixture of transducing phage and recipient cells on rich plates containing tetracycline. Every Tet⁽ʳ⁾ colony that arises will be due to transposition of T-POP from the transduced fragment into the recipient chromosome. Since each colony is the product of a different transposition event, each of the insertion mutants arises independently; no two should be alike. By replica printing, the tetracycline-resistant transductants are transferred by replica printing to lactose indicator plates with and without added tetracycline. Any colony that shows a tetracycline-inducible or tetracycline-repressible phenotype is kept for characterization. P22 will be grown on insertion mutants of interest and used to transduce the T-POP into LT2 to be screened for tetracycline-regulated cell motility on soft agar plates used to detect swimming (and thus expression of flagellar genes).

Procedure 1. Growing P22 Lysates

Solutions

Luria broth (LB): 10 g tryptone, 5 g yeast extract, 5 g NaCl per liter of deionized water

Ex50 salts

50% D-glucose

Sterile saline: 8.5 g NaCl per liter of deionized water

Top agar: 10 g tryptone, 7 g agar per liter of deionized water

P22 broth: 200 ml LB, 2 ml Ex50 salts, 0.8 ml 50% D-glucose, 10^7 to 10^8 plaque-forming units (pfu)/ml P22 transducing phage (P22 HT/*int*)

P22 HT/int Lysate Preparation

Grow a P22-sensitive host strain to saturation in LB. Make serial dilutions of a P22 lysate in sterile saline and plate 0.1 ml of diluted phage with 0.1 ml of cell culture in 3 ml of top agar on a LB agar plate (12 g/l of agar). Pick a single plaque with a Pasteur pipette and inoculate a 1-ml LB-saturated culture of a sensitive strain. Add 4 ml of P22 broth that does not have added P22 and grow with shaking at 37° for 5 h or more. (Lysates left over the weekend will work, but usually an all-day or overnight incubation period is used for convenience.) Titer the resulting lysate and use it to prepare a working stock of P22 broth. For all future lysates, add 4 ml of P22 broth to 1 ml of LB-saturated culture of a sensitive strain and grow 5 to 36 h at 37° with shaking. Pellet cells by centrifugation (10 min at full speed in a table-top centrifuge, or for larger volumes spin 5 min at 8000 rpm in a SS34 rotor). Decant the supernatant into a sterile tube, add $CHCl_3$, and vortex to sterilize. Store at 4°.

MudP22 Lysate Preparation

Grow a strain carrying a MudP22 lysogen and P22 gene 9–expressing plasmid in LB to an OD_{600} of 0.4 to 0.6. Add a 1/100 volume of a 10 mg/ml mitomycin C stock solution (stored at 4°) and grow at 37° with shaking until visible lysis occurs (2 to 4 h). Pellet cells by centrifugation (as above). Decant the supernatant into a sterile tube, add $CHCl_3$, and vortex to sterilize. Store at 4°.

Procedure 2. Transposition of T-POP into the Chromosome of S. enterica*: Isolation of T-POP Insertions Affecting Gene Expression*

Materials

P22 HT/*int* transducing lysates on *S. enterica* strains carrying T-POP insertions in F plasmid DNA (TH3466 [*proAB47*/F′128 (*pro*(+)-*lac*+*zzf-3832*::Tn*10d*Tc[*del-20*] (T-POP1)], TH3467 [*proAB47*/F′128 (*pro*(+)-*lac*+*zzf-3833*::Tn*10d*Tc[*del-25*] (T-POP2)], or TH3468 [*proAB47*/F′128 (*pro*(+)-*lac*+*zzf-3834*::Tn*10d*Tc[*del-20   del-25*] (T-POP3)])

Mu*d*P22 lysate from strain TH3923 [pJS28(Ap^R P22–9+)/F′114(ts) *lac*+ *zzf-20*::Tn*10*[*tet*^{A}::Mu*d*P] *zzf-3823*::Tn*10d*Tc[*del-25*]/*leuA414hs dSBFels2*−].

Recipient strain: pNK(Ap^R Tn*10* transposase, constitutively expressed)/TH2795 (*fliC*::Mu*d*J)

1. Start a 2-ml overnight culture of a *fliC*::Mu*d*J (*fliC-lac*) *Salmonella typhimurium* strain (TH2795) carrying a plasmid (designated pNK) expressing Tn*10* transposase in LB plus ampicillin (Ap). Grow overnight with aeration at 37°.

2. In a sterile tube, mix 1.2 ml of cells from the TH2795 overnight culture with 1.2 ml of phage grown on the T-POP insertion. Some of the phage particles will inject T-POP DNA into your recipient cells. (First do a test cross with different phage stock dilutions, in saline, to obtain 300 to 500 colonies per plate.)

3. Add 0.2 ml of the cell/phage mixture to each of 10 L-Tet plates (Tc). T-POP encodes tetracycline resistance, so in the presence of Tc, only those cells that have inherited the T-POP element will grow. Incubate overnight at 37°.

4. Replica print transduction plates to MacConkey-lactose plates containing tetracycline (Mac-lac tet plates) and Mac-lac plates without tetracycline. Replica print onto the Mac-lac-Tet plate first. Incubate overnight at 30°. The problem here is that tetracycline is bacteriostatic, not bacteriocidal. This means that the background cells that were tetracycline sensitive on the transduction plates will get transferred to the Mac-lac no-tet plate and start growing. *Note*: It is recommended that the cells grow to allow for the visualization of the lactose phenotypes in the presence and absence of tetracycline, but the plates need to be scored before the background grows up and one can no longer distinguish the Tet-resistant colonies printed from the background.

5. Read replica prints, and using a toothpick, pick the potential tet-induced Lac$^+$ or tet-induced Lac$^-$ T-POP insertion mutants, and isolate P22-sensitive mutants using green indicator plates.

6. Prepare P22 transducing lysates on T-POP insertion mutants and use these to transduce the wild-type strain LT2 to Tc^R. Isolate P22-sensitive mutants using green indicator plates. Then check for Tc-induced or Tc-repressed motility by poking colonies into motility plates with and without added tetracycline (15 μg/ml).

Procedure 3. Arbitrary PCR for Sequencing Out of the T-POP Mutants

The procedure used to sequence out of the T-POP ends has been developed from the procedure used by O'Toole and Kolter (1998).

1. Start overnight cultures of the T-POP mutants.

2. Dilute 1:100 of the overnight culture in Millipore water for each sample. Boil the sample for 5 min, then freeze (in dry ice) for 5 min, boil for 5 min, and then freeze for 5 min. Thaw and put on ice. Use 1 to 5 μl for PCR template (see below).

3. Set up a first PCR reaction using the template prepared above and primers Tn10-R1A (to sequence out from the right end of the T-POP) in combination with the arbitrary primers ARB1-A, ARB1-B or ARB1-C (try the three reactions and keep the one that gives the most PCR fragments after the second PCR). The sequence of the primers is shown in Table I. The thermocycling program used is 3 min of denaturation at 95°, followed by 30 cycles of 30-sec denaturation at 95°, 30-sec annealing at 38°, 1.5-min elongation at 72°, and an additional 3 min at 72°. The PCR fragments are purified using the PCR Qiagen kit.

4. A second PCR reaction is conducted using 5 μl of the purified PCR fragment of the first PCR reaction, and primers Tn10–2R and the primer ARB-1 (end sequence of the arbitrary primers). The thermocycling program used is 3-min denaturation at 95°, followed by 30 cycles of 15-sec

TABLE I

SEQUENCE OF PRIMERS FOR ARBITRARY PCR FOR SEQUENCING OUT FROM THE ENDS OF THE T-POP

Primer name	Sequence
Arbitrary primer PCR	
ARB-1A	5′ GGC CAG CGA GCT AAC GAG ACN NNN GTT GC-3′
ARB-1B	5′ GGC CAG CGA GCT AAC GAG ACN NNN GAT AT-3′
ARB-1C	5′ GGC CAG CGA GCT AAC GAG ACN NNN AGT AC-3′
ARB-1	5′ GGC CAG CGA GCT AAC GAG AC-3′
Right end T-POP, TN10, TN10dTc	
Tn10R1A	5′AAT TGC TGC TTA TAA CAG GCA CTG-3′
TN10-2R	5′ ACC TTT GGT CAC CAA AGC TTT-3′

denaturation at 95°, 30-sec annealing at 56°, 1.5-min elongation at 72°, and an additional 3 min at 72°. The PCR fragments are purified using the PCR Qiagen purification kit.

5. The big dye sequence is conducted using 200 ng of the second-reaction PCR product and 4 pmol of the primer Tn10–2R. The big dye thermocycling reaction is 40 cycles of the following steps: 96° for 30 sec, 55° for 15 sec, and 60° for 4 min.

References

Kleckner, N., Bender, J., and Gottesman, S. (1991). Uses of transposons with emphasis on Tn*10*. *Methods Enzymol.* **204,** 139–180.

O'Toole, G. A., and Kolter, R. (1998). Flagellar and twitching motility are necessary for *Pseudomonas aeruginosa* biofilm development. *Mol. Microbiol.* **28,** 449–461.

Rappleye, C. A., and Roth, J. R. (1997). A Tn*10* derivative (T-POP) for isolation of insertions with conditional (tetracycline-dependent) phenotypes. *J. Bacteriol.* **179,** 5827–5834.

Section III

Phage

[15]   Recombineering: *In Vivo* Genetic Engineering in *E. coli, S. enterica*, and Beyond

By James A. Sawitzke, Lynn C. Thomason, Nina Costantino,
Mikhail Bubunenko, Simanti Datta, and Donald L. Court

Abstract

"Recombineering," *in vivo* genetic engineering with short DNA homologies, is changing how constructs are made. The methods are simple, precise, efficient, rapid, and inexpensive. Complicated genetic constructs that can be difficult or even impossible to make with *in vitro* genetic engineering can be created in days with recombineering. DNA molecules that are too large to manipulate with classical techniques are amenable to recombineering. This technology utilizes the phage λ homologous recombination functions, proteins that can efficiently catalyze recombination between short homologies. Recombineering can be accomplished with linear PCR products or even single-stranded oligos. In this chapter we discuss methods of and ways to use recombineering.

Introduction

What Is Recombineering?

In vivo genetic engineering using the bacteriophage lambda (λ) recombination proteins and short DNA homologies has been termed "recombineering" (**recombi**nation-mediated genetic engi**neering**) (Ellis *et al.*, 2001) and is the subject of this chapter.

Genetic engineering has been instrumental in revolutionizing studies in molecular biology for over 30 years since the discovery of restriction enzymes. *Escherichia coli* has been the standard host used to recover the products of this *in vitro* genetic engineering. Since the late 1990s, however, new *in vivo* technologies have emerged that greatly simplify, accelerate, and expand genetic engineering in *E. coli, Salmonella enterica*, and other organisms. Now, within a week a researcher can modify any nucleotide(s) of choice in almost any manner. Further, these genetic engineering technologies do not rely on *in vitro* reactions carried out by restriction enzymes and DNA ligase. Instead, they utilize the bacteriophage λ homologous recombination proteins collectively called "Red" to directly modify DNA within a bacterial cell. Importantly, the Red proteins require only ∼50 bases

METHODS IN ENZYMOLOGY, VOL. 421   0076-6879/07 $35.00
                                                   DOI: 10.1016/S0076-6879(06)21015-2

of homology to catalyze efficient recombination. These homologies are small enough that they can be provided by synthetic oligonucleotides.

Red Proteins and Properties

Homologous recombination is the process whereby segments of DNA are exchanged between two DNA molecules through regions of identical DNA sequence, the end result being new combinations of genetic material. Generalized recombination catalyzed by the *E. coli* recombination proteins occurs when there are about 100 base pairs of homology for exchange and becomes more efficient with longer homologies (Shen and Huang, 1986; Watt *et al.*, 1985).

Normally, linear DNA introduced into *E. coli* is degraded by the powerful RecBCD nuclease. Although *in vivo* genetic engineering systems have been previously attempted (for a review of other systems, see Court *et al.*, 2002), none have been fully satisfactory. In contrast, the Red proteins of phage λ and the RecET proteins of the cryptic *rac* prophage have properties that allow recombination of a linear, modifying DNA containing short (~50 bp) homologies with appropriate target sequences, thereby allowing rapid and efficient genetic engineering (Muyrers *et al.*, 1999, 2000; Yu *et al.*, 2000; Zhang *et al.*, 1998, 2000). Other similar systems will also undoubtedly be developed (Poteete, 2001; Poteete and Fenton, 1993; Vellani and Myers, 2003); however, in this review we concentrate on the λ Red system, Exo, Beta, and Gam.

The λ Gam protein inhibits the RecBCD and SbcCD nuclease activities, preserving linear DNA and thereby allowing it to be used as a substrate for recombination (Chalker *et al.*, 1988; Gibson *et al.*, 1992; Karu *et al.*, 1975; Kulkarni and Stahl, 1989; Murphy, 1991). Linear DNA is required for Red-mediated recombination (Stahl *et al.*, 1985; Thaler *et al.*, 1987a, 1987b). This can be either a linear double-strand DNA (dsDNA) generated by PCR or a short single-stranded DNA (ssDNA) oligonucleotide (oligo) carrying homology to the target (Court *et al.*, 2002).

The λ Exo protein, a dsDNA-dependent exonuclease, processes linear dsDNA. Exo requires a dsDNA end to bind and remains bound to one strand while degrading the other in a 5'-3' direction (Carter and Radding, 1971; Cassuto and Radding, 1971; Cassuto *et al.*, 1971). This results in dsDNA with a 3' ssDNA overhang, the substrate required for the Beta protein to bind. Exo is required only for recombineering with dsDNA substrates (Ellis *et al.*, 2001; Yu *et al.*, 2000).

λ Beta is a ssDNA-binding protein that can promote the annealing of complementary DNA strands. Beta can bind stably to ssDNA greater than 35 nucleotides (Mythili *et al.*, 1996) and protect the DNA from single-strand nuclease attack (Karakousis *et al.*, 1998; Muniyappa and Radding, 1986). Beta is the only known λ function required for recombineering with ssDNA oligos (Ellis *et al.*, 2001).

Recombineering with linear dsDNA requires all three Red proteins. Gam is needed to protect the linear substrate. Since Beta and Exo form a complex (Radding *et al.*, 1971), it is reasonable to suggest that as Exo degrades a chain of dsDNA, Beta binds to the newly formed ssDNA (Karakousis *et al.*, 1998; Li *et al.*, 1998). However, Beta alone is sufficient for recombination with ssDNA substrates (Ellis *et al.*, 2001; Yu *et al.*, 2000).

Expression of Red Proteins from a Defective Prophage

The Red proteins are encoded by the *gam*, *bet*, and *exo* genes located next to each other in the *pL* operon of λ. The timing and level of expression of these genes is of critical importance for the highest recombineering efficiencies. Prolonged expression of Gam can lead to plasmid instability (Murphy, 1991; Silberstein and Cohen, 1987; Silberstein *et al.*, 1990) and toxic effects to the cell (Friedman and Hays, 1986; Sergueev *et al.*, 2001). Inappropriate expression of Exo and Beta can lead to unwanted rearrangements, which is especially problematic in working with eukaryotic DNA cloned into BACs.

For ease of movement between strains, several labs have cloned various combinations of the Red genes on plasmids under the control of heterologous promoters (Datsenko and Wanner, 2000; Muyrers *et al.*, 1999, 2000; Zhang *et al.*, 1998, 2000). Although these systems have been effectively used for recombineering, plasmid-borne systems can be prone to inappropriate expression problems. More recently developed plasmid systems have abrogated some of these problems (see "Prophage-Containing Recombineering Plasmids" section).

Our laboratory has developed and utilized a λ prophage for expression of the Red genes (Court *et al.*, 2002; Ellis *et al.*, 2001; Yu *et al.*, 2000). The prophage is defective in that it has been deleted for the lysis, DNA replication, and structural genes of the phage, but retains the critical features of transcriptional control and importantly, the Red functions (Fig. 1A). With this prophage, the Red genes are expressed from the *pL* operon under the control of the temperature-sensitive repressor, CI857. Thus, when the cells are at a low temperature ($<37°$), the CI857 repressor is active and there is no expression of the Red genes except in a rare subpopulation of spontaneously induced cells. After a brief temperature upshift to $42°$, the CI857 repressor denatures, allowing transcription from *pL* and thereby Red expression. Upon shifting back to low temperature, CI857 renatures and again completely blocks transcription of *pL*. Thus, Red functions are

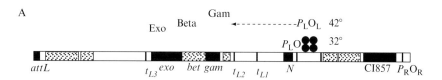

F<small>IG</small>. 1. Diagram of the defective prophage used for recombineering. (A) Standard defective prophage originally described in Yu *et al.* (2000). The *red* genes—*exo*, *bet*, and *gam*—are under control of the temperature-sensitive repressor, CI857. Transcription of the *red* genes (beyond t_{L1}) requires the N protein. (B) The minimal prophage as described in Datta *et al.* (2006). Transcriptional terminators t_{L1} and t_{L2} as well as the N gene have been deleted. The minimal prophage is no longer dependent on N protein but still is regulated by CI857. In both cases, at temperatures less than 34°, CI repressor (filled circles) binds the operators and prevents transcription of the p_L operon. At 42°, the temperature-sensitive CI857 repressor denatures and thus allows transcription of the *red* genes.

available for a short but sufficient time to recombine the sequences of interest and then they are removed to minimize extraneous events. Gam is extremely toxic to cells, but this short pulse of expression does not interfere with cell viability (Sergueev *et al.*, 2001).

In this review, we focus on using recombineering to manipulate DNA on the bacterial chromosome, plasmids, or phage. However, recombineering is just as useful to modify BACs containing DNA from other organisms for functional genomic studies (Copeland *et al.*, 2001; Lee *et al.*, 2001; Muyrers *et al.*, 1999; Swaminathan *et al.*, 2001; Warming *et al.*, 2005). Discussion of the mechanism(s) of recombineering can be found elsewhere (Costantino and Court, 2003; Court *et al.*, 2002; Ellis *et al.*, 2001).

Methods

Standard Recombineering Protocol

The steps for executing the standard recombineering protocol in *E. coli* or *S. enterica* include: (1) preparation of electrocompetent cells that contain the λ recombination proteins needed for recombineering, (2) transformation of those cells with the DNA substrate using electroporation, (3) outgrowth, (4) selection or screening for the chosen genetic change, (5) confirmation of the genetic alteration, and (6) elimination of the λ stuff.

The following protocol outlines the procedure that we have found to produce the most consistent results. Some parameters have been optimized while others have not (Yu *et al.*, 2000, 2003). Any deviation from this protocol may produce less than satisfactory results, but modifications may prove necessary in other organisms.

Preparation of Electrocompetent and Recombineering-Proficient Cells

The first step is to produce cells that are competent for both the uptake of DNA and for recombineering. With our standard prophage expression system where the cells contain the λ *red* genes under CI857 control, a 5-ml overnight culture is grown in Luria broth (LB) at 30 to 32°. This culture is then diluted at least 70-fold (0.5 ml of overnight culture into 35 ml of fresh LB) and grown in a 125-ml baffled flask with shaking (200 rpm) at 32° until the OD_{600} is 0.4 to 0.5. Fifteen milliliters of culture are then rapidly shifted to 42° and incubated with shaking (200 rpm) for 15 min to induce production of the Red proteins. The rest of the cells remain at 32° (the uninduced control). After 15 min, all flasks are placed in an ice-water bath and swirled to rapidly cool them. Flasks are swirled intermittently in the ice bath for 5 to 10 min until the cultures are completely chilled. The cells are pelleted by centrifugation at $4600 \times g$ (6700 rpm in a Sorvall SA-600 rotor) for 7 min in a 4° centrifuge. The supernatant is decanted or aspirated, and the cells are gently suspended with 1 milliliter of ice-cold sterile distilled H_2O using a large disposable pipette tip or gentle shaking. A vortex must not be used for this or subsequent steps as cells in H_2O are fragile. After the cells are suspended, an additional 30 ml of ice-cold sterile distilled water is added to each tube and gently rocked to mix before pelleting again at $4600 \times g$ for 7 min. The pellet will be very loose and great care must be taken not to lose the cells while decanting the supernatant. Again the pellet is gently suspended with 1-ml ice-cold distilled H_2O. The cells are then transferred to a chilled microfuge tube and pelleted in a 4° microfuge for ~30 sec at maximum speed. Finally, each preparation of cells is suspended in 200 μl of ice-cold distilled H_2O and kept on ice until electroporation. This should be enough cells for four or five electroporations. We always use freshly prepared electrocompetent cells for the highest efficiencies, but cells can be frozen at –80° in 12% glycerol for future recombineering, albeit at a lower efficiency (Yu *et al.*, 2000).

Transformation by Electroporation

Once the cells are competent, the DNA substrate is introduced by electroporation. We use the standard conditions recommended for *E. coli* and Salmonella in a Bio-Rad electroporator: 1.8 kV with 0.1-cm cuvettes that have been chilled on ice. Other conditions have not been tested thoroughly by our laboratory. We typically mix 100 to 300 ng of salt-free PCR product (see preparation of linear DNA) or 50 to 100 ng of salt-free ssDNA (oligos) with 50 μl of electro-competent cells. They can be mixed in either a cold microfuge tube and then moved to the cuvette, or mixed directly in the electroporation cuvette with similar results. Important controls include induced cells with no DNA and uninduced cells with DNA. Optimal electroporations give a time constant of more than 5.0 msec. Lower time constants may produce recombinants but at a lower efficiency, and may reduce total cell viability. Immediately after electroporation, 1 ml of LB is added to the electroporation cuvette, and cells are transferred to a sterile culture tube. Subsequent steps depend on the specifics of the desired recombination event.

Outgrowth

Once LB has been added to the electroporated cells, a minimum 30-min incubation at 32° is necessary to allow their recovery from electroporation. Several outgrowth options are available; the appropriate one depends on the type of recombinants generated and the method being used to identify recombinants. In general, the options are to dilute and spread the dilutions on agar plates after the 30-min outgrowth, or to incubate longer and grow the electroporation mixture in LB before dilution and plating. In the first case, each electroporated cell is plated before significant cell division occurs, and in the second case, the electroporated cells grow and divide before plating.

At the time of recombination, there are several replicating copies of the bacterial chromosome (four to eight), but recombination is restricted in most instances to one of these and, in the case of oligonucleotide recombinants, to one strand of one copy (Costantino and Court, 2003). Thus, during further growth of these cells (either on plates or in LB), the DNA copies present at recombination segregate from one another, separating recombined from unrecombined DNA copies. If cells are spread on agar before outgrowth, recombinant colonies that form will be a mixture of recombinant and parental cells. If sufficient time is allowed for outgrowth in liquid culture, each colony will be relatively pure, but the frequency of recombinant colonies will be reduced by the outgrowth and segregation process. This dilution effect could be as much as 4- to 16-fold for *E. coli* growing in LB because of the multiple replication forks and DNA copies present at the time of electroporation and recombination (Sergueev *et al.*, 2002).

Outgrowth before plating is critical for finding recombinants in certain situations. For example, when a drug-resistance cassette is used for targeting, recombinants are selected in the presence of the drug. In this situation, the recombinant cassette must be expressed before the cell carrying it is challenged with the drug. Usually 2 to 3 h of outgrowth in the absence of drug selection are required for sufficient expression. As a different example, when a gene that makes a conditionally toxic product to the cell is targeted for replacement by recombination, then complete segregation should be allowed so that only a pure recombinant cell (i.e., one that does not contain the toxic gene) remains (to avoid toxicity on selection). Examples of this are the counter-selected genes such as *sacB*, *galK*, and *thyA*, which will be described later. In this case, a longer outgrowth in liquid media should be allowed to generate recombinant cells free of the gene and its toxic product.

Plating cells soon after electroporation reduces the number of colonies that need to be screened when nonselective procedures are used to find recombination. Once recombinant colonies are found, however, the recombinant cells within the colony must be purified away from the parental segregants, which are also present.

Selection or Screening for Mutants

When cells are ready for dilution and plating, tenfold stepwise dilutions should be made in TMG, minimal salts, or similar osmotically balanced medium (Arber *et al.*, 1983; Sambrook and Russell, 2001). Luria broth may be used for dilutions if selection is for a drug resistance. The appropriate dilution and plates to use for selecting/screening for recombinants depends on the specifics of the recombineering being performed. In initial experiments, a wide range of dilutions should be plated for both selection of recombinants and determination of cell viability. For example, if a PCR product was used to insert a drug cassette, then we optimally see 10^3 to 10^4 recombinants per 10^8 viable cells (Table I). If, however, an oligo (ssDNA)

TABLE I

A Comparison of Recombineering Efficiencies with Various Substrates

	Number of Recombinants/10^8 Viable Cells			
		Oligo Repair with Lagging Strand[a]		
Strain	dsDNA[b]	T/C[c]	C/C	Multibase mismatch[d]
Wild-type	$\sim10^4$	$\sim10^5$	$\sim10^7$	$\sim10^7$
*mut*S	$\sim10^4$	$\sim10^7$	$\sim10^7$	$\sim10^7$

[a] Using the leading strand, recombination is up to 30-fold reduced as compared to the lagging strand.
[b] For example, replacing the *galK* gene with a drug cassette.
[c] Or any mispair other than a C/C.
[d] Four or more mismatches in a row.

is used for recombineering a point mutation, the frequency of recombination is routinely 10^5 per 10^8 viable cells, and under some conditions may be as high as 25% of the total viable cells (Table I) (Costantino and Court, 2003). In the strains that we use, we find 10^7 to 10^8 viable cells per milliliter after electroporation and a 2-h outgrowth. In some strains we see up to a 10-fold reduction in viability after electroporation. It is important to verify total viable cells to ensure there are enough cells to isolate recombinants. To determine the total cells that survive electroporation, dilutions are plated nonselectively on L plates and incubated at $\leq 34°$.

If a high level of recombination is expected ($>10^5/10^8$ viable), cells can be plated nonselectively on L plates and recombinants screened for by checking individual colonies for the desired phenotype or genotype. For example, if ssDNA was used to recombineer a new restriction site into a gene, a diagnostic PCR fragment followed by restriction analysis can be used to identify the recombinant colonies. Single base changes can also be detected by the mismatch amplification assay-PCR (MAMA-PCR) method (Cha et al., 1992; Swaminathan et al., 2001). Another method to screen for nonselected recombinants is colony hybridization of cells (L. C. Thomason, et al., unpublished results, 2005b). For this method, the sequence inserted by recombineering must be unique to the recombinant so it can be used as a probe. Finally, in some cases, it is possible to detect recombinants directly on nonselective plates. For example, if the recombinant produces altered colony morphology or a slow-growth phenotype, these can be detected directly by looking for that minority class of colonies (Thomason and Sawitzke, unpublished results).

As an alternative to screening nonselected colonies, a two-step selective protocol can be used to modify a region of interest. First, the targeted region is replaced by a dual selection cassette such as cat-sacB (see "Selection/Counter-Selection for Gene mutation, Replacement, and Fusion" section), then an oligo (or PCR product) containing the mutations can be introduced in the second step. With this method, there is selection for both steps so that no screening is required. This protocol is useful for making numerous site-specific mutations in a region of interest.

Confirming Mutations

Candidate recombinants must be purified by streaking out for single colonies on the appropriate plates before further testing. Once recombinant candidates have been purified, the desired changes can be confirmed by PCR analysis, restriction analysis, and DNA sequencing. Sequence analysis will also confirm that no extraneous changes were made. It is known that inadvertent changes can arise because of errors introduced during oligosynthesis (Oppenheim et al., 2004).

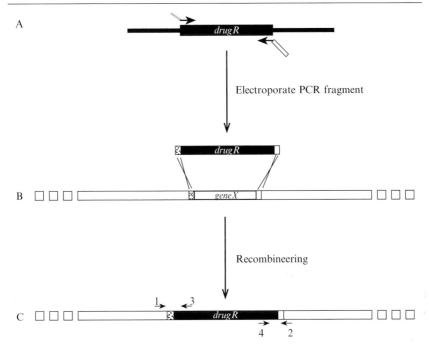

FIG. 2. Using recombineering to replace a gene with a drug-resistance cassette. (A) A pair of hybrid primers that contain at their 5′ end, ~50 bases of homology to the intended target, and at their 3′ end, sequence for priming a template for a drug-resistance (drugR) cassette (Table III). PCR using these primers and the proper template produces the linear substrate with the drugR flanked by 50-bp homologies. The primer design determines precisely where the drug cassette will insert. In this example, we fully replace "geneX" with a drug cassette using homologies that flank geneX. (B) The drugR fragment is electroporated into Red-induced cells where recombineering takes place. (C) Drug-resistant clones are checked for gene replacement by PCR analysis. PCR using primers 1 and 3, 2 and 4, and 1 and 2 should yield products of predicted sizes.

For an antibiotic cassette or other insertion, PCR can be used to confirm its location. Two primers, internal to the insertion, should be designed pointing out towards each end of the insert to be paired with primers flanking the site of insertion (see Fig. 2C and legend). Predicted fragments from all the various primer pairs should be checked (Yu *et al.*, 2000). Sequencing can be done to fully verify all junctions if necessary.

Elimination of the λ Stuff

After recombineering, in many cases it is desirable or necessary to remove the *red* (and other λ) genes. This may be accomplished in several ways, and the choice depends on the details of the experiment and which recombineering system is being used. In general, the *red* genes can be removed from the strain in which the recombineering took place, or alternatively, the newly constructed recombinant can be moved to a clean genetic background. For genetic experiments, the latter is usually preferable, especially if a mismatch-repair mutant strain was used.

If the altered DNA resides on a plasmid or BAC, then the newly made construct will often be moved away from the recombineering genes during the course of the protocol by plasmid isolation and re-transformation into a nonrecombineering host (see "Recombineering on a Plasmid" section, and Warming et al., 2005).

If the new construct resides in the chromosome and has a selectable phenotype (e.g., drug resistance or auxotrophy), generalized transduction using phage P1 (P22 in S. enterica) can be used to move it to a clean genetic environment, away from the recombineering strain. Using generalized transduction to move a point mutation on the chromosome, especially one without a selectable or easy-to-screen phenotype, can be difficult to accomplish. In such cases, it may be necessary or at least easier to remove the recombineering system from that strain. If the defective prophage was used for recombineering, then it can be removed either by generalized transduction (e.g., use linked nadA::Tn10) or by recombineering a PCR fragment of the wild-type attB bio region made from a nonlysogen to replace the prophage (Yu et al., 2000). You can select for growth on minimal medium without biotin at 42° since the prophage makes the strain temperature sensitive and a biotin auxotroph.

Some of the prophage-containing recombineering plasmids have a temperature-sensitive origin of replication (Table II), and a temperature shift will encourage loss of the plasmid (Datta et al., 2006). Plasmid loss is accomplished by diluting an overnight cell culture containing the temperature-sensitive plasmid 1000-fold in LB and growing at 37° for more than 4 h. Dilutions are then plated on L plates at 32°. After this regimen, nearly 100% of tested colonies have lost the plasmid.

TABLE II
RED-PRODUCING PLASMIDS

Plasmid[a]	Origin	Drug resistance
pSIM5	pSC101 repA[ts]	Chloramphenicol
pSIM6	pSC101 repA[ts]	Ampicillin
pSIM7	pBBR1	Chloramphenicol
pSIM8	pBBR1	Ampicillin
pSIM9	pRK2 trfA[ts]	Chloramphenicol
pSIM18	pSC101 repA[ts]	Hygromycin
pSIM19	pSC101 repA[ts]	Spectinomycin

[a] Plasmids are further described in Datta et al., 2006.

Preparation of Linear DNA for Recombineering

Linear DNA that is either single- or double-stranded is needed for recombineering. Whether you should use ss- or ds-DNA depends on the details of the construct being made.

Oligo Design for ssDNA Recombineering

For ssDNA recombineering, we order salt-free oligos with no further purification. In some cases, gel purification can be used to reduce unwanted base deletion mutations introduced during oligo synthesis (Oppenheim *et al.*, 2004). If there is a selection for function, then most of these unwanted mutations in the oligo will be selected against. The oligo is reconstituted at a concentration of 1 nmol/μl in Tris EDTA(TE) and stored at $-20°$. Multiple freeze/thaw cycles are avoided by making working stock aliquots at a final concentration of 10 pmol/μl in dH$_2$0. Use 0.5 μl of this working stock for 50 μl of electro-competent cells. We use 70 base oligos for recombineering. Base changes should be centered in the oligo as much as possible, although anywhere within the "middle" 20 bases of a 70-base oligo give similar frequencies of recombinants (Costantino and Court, unpublished results).

For a given target, there are two complementary ssDNA oligos, either one of which can be used for recombineering. One corresponds to the DNA strand that is replicated as the "leading strand" and the other to the "lagging strand." The lagging strand oligo corresponds in sequence to Okazaki fragments. The efficiency of recombination is up to 30-fold higher with the oligo that corresponds to the lagging strand (Costantino and Court, 2003; Ellis *et al.*, 2001). These data help support the model that Beta anneals the ssDNA oligo at the DNA replication fork (Court *et al.*, 2002; Ellis *et al.*, 2001). Thus, for ssDNA recombineering, the oligo of choice is the one that corresponds to the lagging strand sequence.

Preparing Linear dsDNA

If linear dsDNA is the substrate for recombineering, PCR is normally used to generate this substrate. We use standard reaction conditions with a high-fidelity PCR kit. Each ~70 base salt-free primer contains two parts (Yu *et al.*, 2000)—the 5′ ends contain the ~50 bases of homology to the target, whereas the 3′ end of the oligo primes the DNA to be inserted (Yu *et al.*, 2000). Thus, the precise join point of the final recombinant product is defined by the oligo design (Fig. 2). When creating deletions, gene replacements, or fusion proteins with recombineering, it is important to keep polarity in mind. An out-of-frame replacement can potentially eliminate expression of downstream genes causing unintended phenotypes.

The PCR-generated targeting DNA often contains a drug-resistance marker flanked by homology sequences, but it can contain any sequence that can be selected or screened for. Recombineering to insert or remove a large heterology is less efficient than creating a single base change (Table I), so a direct selection or a two-step selection/counter-selection (see below) should be used when possible. Table III details the primers we use for amplifying drug cassettes with their promoters and transcription terminators. They have been chosen to allow efficient PCR synthesis, and ultimately, expression of the drug cassette. The PCR products are purified with a commercially available PCR cleanup kit before recombineering.

The method used for PCR amplification can have dramatic effects on the experimental results. Often the template for the PCR is a plasmid from which drug-resistance and other cassettes are amplified. It is important to use the least amount of plasmid DNA possible for the reaction. Template plasmid DNA still present during electroporation will give rise to drug-resistant colonies because transformation of supercoiled plasmid is very efficient. Plasmid DNA can be greatly reduced after the PCR reaction by digesting with DpnI, which cuts methylated DNA but not the unmethylated PCR products. Transformation of uninduced cells with the linear vector mix will give an estimate of the amount of uncut plasmid template still present in the preparation. This is an important control.

Because of the problems caused in getting rid of plasmid DNA, nonplasmid templates may be preferred. For example, cassettes already cloned into the bacterial chromosome can be amplified. Alternatively, PCR amplified cassettes can be maintained as stock DNA templates for subsequent amplification. Care must be taken if the template for PCR is also a PCR product, since serial amplifications will cause mutations to accumulate in the PCR products, thus resulting in problems. We have seen the *sacB* gene become less sensitive to sucrose as a result of repeated amplifications (Thomason, unpublished results). Therefore, make a stock template once from an original source. Once it is used up, make a new stock from the original source.

Maximizing Recombination

Methyl-directed mismatch repair (MMR) reduces recombination frequencies (Costantino and Court, 2003). The MMR system recognizes and repairs base pair mismatches and small (1 to 3 bp) deletions, but not larger heterologies. In the absence of MMR activity, recombination frequencies can be increased. The frequency of recombineering to insert or remove a large heterology is not affected by mismatch repair.

Methyl-Directed Mismatch Repair Mutants

In *E. coli*, the MMR system includes, among other functions, MutH, MutL, MutS, the UvrD helicase, and the Dam methylase. Cells containing

TABLE III
Primer Pairs for Amplifying Cassettes[a]

Drug cassette	Potential template sources[a]	Primer pair
Ampicillin	pBR322 (New England Biolabs) and derivatives	5' CATTCAAATATGTATCCGCTC 5' AGAGTTGGTAGCTCTTGATC
Kanamycin	pBBR1MCS-2 (Kovach et al., 1994), Tn5 (Ahmed and Podemski, 1995) Note: this is not the same kanamycin gene as in Tn903.	5' TATGGACAGCAAGCGAACCG 5' TCAGAAGAACTCGTCAAGAAG
Chloramphenicol	pACYC184 (New England Biolabs)	5' TGTGACGGAAGATCACTTCG 5' ACCAGCAATAGACATAAGCG
Tetracycline	Tn10 (Hillen and Schollmeier, 1983) Note: this is not the same tetracycline gene as in pBR322 or pACYC184	5' CAAGAGGGTCATTATATTTCG 5' ACTCGACATCTTGGTTACCG
Spectinomycin	pBBR1MCS-5 (Kovach et al., 1994), DH5αPRO (Clontech)	5' ACCGTGGAAACGGATGAAGGC 5' AGGGCTTATTATGCACGCTTAA
cat-sacB cassette	pK04/pEL04 (Lee et al., 2001)	5' TGTGACGGAAGATCACTTCG 5' ATCAAAGGGAAAACTGTCCATAT
PCR fragment to remove prophage	E. coli	5' GAGGTACCAGGCGCGGTTTGATC 5' CTCCGGTCTTAATCGACAGCAAC

[a] We often grow an overnight of cells containing the desired drug-resistance template in the chromosome; 2 μl of this overnight is an excellent template for PCR. We have listed some commonly found sources of these sequences, but others may be suitable. As multiple versions of drug-resistance cassettes are available (as noted above), caution must be used to be certain that these primers will prime your template.

Notes: All primers included in this table are designed so that the PCR product will contain a promoter (if appropriate) for the drug-resistance gene. All cassettes except for the kanamycin gene also contain a transcription terminator. We are currently engineering a terminator for the kanamycin cassette. Using other priming oligos that are not shown here, a PCR product can be generated to replace a gene from its start to stop codons with a drug-resistance cassette. Using other priming oligos that are not shown here, a PCR product can be generated to replace a gene from its start to stop codons, thus producing the drug-resistant recombinant with the gene's native regulation.

a mutation that eliminates any of these functions exhibit increased levels of recombination with ssDNA, given that the recombinants are no longer removed by the MMR system (Costantino and Court, 2003; Li et al., 2003). More than a 100-fold increase in recombination can be achieved by eliminating the MMR system when changing a single base (Table I). This increase allows up to 25% of the cells surviving electroporation to become recombinants when a lagging strand oligo is used, making screening for recombinants easy. The drawback to this method is that MMR-deficient strains are mutagenic, causing the frequency of extraneous mutations to be increased.

C/C Mismatch

With careful design, high levels of recombineering can be achieved in strains that are wild-type (WT) for mismatch repair (Costantino and Court, 2003). This is possible because some mismatches are poorly corrected by the MMR system. The hierarchy of repair from poorest to most efficiently repaired is C/C < A/G, T/C, T/T < G/G, A/A, A/C, G/T (Dohet et al., 1986; Su et al., 1988). If the recombining oligo creates a C/C mismatch when annealed to the target sequence, this mismatch is not recognized by the MMR system and is not repaired. In practical terms, this means that any G can be efficiently changed to a C. In fact, a C/C mispair within 6 bp upstream or downstream of a second desired change prevents the second change from being repaired (N. Costantino and D. Court, unpublished results). Thus, generating C/C mismatches allows high levels of recombineering at many positions without the negative side effect of the strain being mutagenic.

Other Means of Maximizing Recombination

Another method to evade the MMR system while recombineering is to design the oligo with multiple adjacent base changes. With careful design the additional changes can introduce or remove a restriction site that will aid confirmation. Using this trick, a single point mutation can be made in two steps with high levels of recombination in both steps (Yang and Sharan, 2003). With the first event, four to six changes are made that cover the mutational site of interest. Next, a second oligo recombination event can be used to change the sequence back to WT except for the desired point mutation.

Finally, the MMR system can be inhibited temporarily by a dominant negative allele of the *mutS* gene (Haber and Walker, 1991) or by addition of 2-aminopurine (2-AP) (Costantino and Court, 2003). Incubation of cells for 3 h with 75 μg/ml 2-AP increased the level of recombination, but not to that obtained with the complete absence of mismatch repair. Thus, 2-AP can be used to increase recombination frequencies with limited general mutagenesis of the cells.

Genetic Manipulations

Several other useful genetic tricks are available that facilitate the manipulation of DNA with recombineering. With this toolkit, nearly any construct can be made efficiently and seamlessly.

Selection/Counter-Selection for Gene Mutation, Replacement, and Fusion

Another two-step protocol is frequently used to make changes for which there is no selection. This method is useful to make a protein fusion that has no obvious phenotype, to mutagenize a region, or alter a specific base and leave no other changes. In the first step, dual selection cassettes containing both selectable and counter-selectable markers are recombineered into the target location. At this first step, selection is used to insert the markers near a base or region to be changed. In the second step, counter-selection is used to replace the dual selection cassette with the final DNA construct.

We routinely use the *cat-sacB* cassette (Ellis *et al.*, 2001; Thomason *et al.*, 2005a) with an initial selection for chloramphenicol resistance in the first round of recombineering, and a final selection *sacB* in the second round. The *sacB* gene makes *E. coli* sensitive to sucrose; thus, plates containing sucrose (see "Media") can be used to select against cells containing this gene (Gay *et al.*, 1985). After insertion of the cassette by recombineering and selection for chloramphenicol-resistant recombinants, several isolates should be purified and tested for sucrose sensitivity. We have found instances when the expression of the *sacB* cassette is affected by its orientation at the target (L. C. Thomason *et al.*, unpublished results). Thus, at some loci, both orientations may need to be tried to ensure a strong counter-selection. A sucrose-sensitive isolate is chosen for the second round of recombineering, from which sucrose-resistant colonies are selected and screened to confirm that they are chloramphenicol sensitive and true recombinants. Those that are still resistant to chloramphenicol may have a spontaneous mutation in the *sacB* gene (normally found at a frequency of 1 in 10^4), and thus are "false positives." If recombination conditions have been optimized, the number of chloramphenicol-sensitive recombinants should be greater than these chloramphenicol-resistant false positives.

Recently, *galK* and *thyA* have been developed for the same purpose as *cat-sacB* (Warming *et al.*, 2005; Wong *et al.*, 2005); however, in these cases, either *galK* or *thyA* is used for both selections. To use *galK* as a dual selection cassette, the recombineering takes place in cells that are deleted for the *galK* gene, and thus are unable to utilize galactose as a sole carbon source. In the first step, recombineering inserts the *galK* gene, allowing growth on minimal galactose agar. The *galK* gene product, galactokinase, also effectively catalyzes the phosphorylation of the galactose analog, 2-deoxy-galactose (DOG), leading to a toxic buildup of 2-deoxy-galactose-1-phosphate (Alper and Ames, 1975). Thus, the second round of recombineering with the *galK* system is selection against *galK* on agar containing DOG (see "Media" section).

When *thyA* is used as the dual selection cassette, the cells must be deleted for *thyA* (Wong *et al.*, 2005). Cells containing a *thyA* deletion are unable to grow on minimal medium in the absence of thymine. Thus, in the first recombineering step, *thyA* is inserted in the target sequence of cells that contain a *thyA* deletion, selecting for growth on minimal medium. Cells containing a functional *thyA* gene, however, are sensitive to trimethoprim in the presence of thymine, which is the basis for the counter-selection in the second recombineering event.

There is one minor change to the "basic protocol" for the second recombineering event when using a selection/counter-selection. The electroporated cells should be suspended in a final volume of 10 ml of LB and incubated with aeration at $\leq 34°$ for at least 3 to 4 h, and preferably overnight. The longer outgrowth allows for complete segregation of recombinant chromosomes that no longer contain the toxic counter-selectable marker. The presence of a sister chromosome with an intact counter-selectable marker will prevent growth of the cell even though one chromosome is recombinant. We note, however, the standard recombineering protocol that includes outgrowth for 3 hr in 1 ml of broth does produce some recombinants.

All of the dual selection systems have strengths and weaknesses. The *cat-sacB* product is large (~ 3 kb), and thus the PCR product can be more difficult to make than the single gene (*galK* and *thyA*) systems. The *cat-sacB* dual cassette system will work in any strain and has the added advantage that loss of the *cat* cassette can easily be screened. In contrast, the *galK* or *thyA* systems work only in strains lacking these genes, and PCR must be used to distinguish true recombinants from spontaneous mutations. Note that the *cat* gene can be replaced by another drug-resistance marker in the *cat-sacB* dual selection cassette.

Duplications

Recombineering can be used to identify duplications, which are tandem diploid regions often being multiple kilobases in size. Duplications naturally occur and exist for any region at frequencies from 10^{-4} to 10^{-2} in a culture (Haack and Roth, 1995). Cells with such duplications can be identified by engineering a gene replacement with a selectable drug cassette in which the gene being replaced is either essential or is conditionally essential (Yu *et al.*, 2000). The duplication is stabilized by maintaining simultaneous selection for the essential gene and the drug cassette. If one targets genes in the chromosome, two classes of recombinants are found based on frequencies alone. Replacement of a nonessential gene is straight-forward and occurs at high efficiency, whereas replacement of an essential gene occurs but is found at much reduced frequency ($<100/10^8$ viable). Such rare recombinants contain large duplications with a second WT copy of the essential gene present. PCR analysis using primers that flank the targeted essential gene is useful for identifying the duplication, as two products will be seen corresponding to the essential gene and the modified copy (M. Bubunenko, unpublished results).

Recombineering can also be used to engineer duplication of a defined region by designing the linear substrate with the appropriate homologies (Sawitzke, unpublished results). This technique is described in Slechta *et al.* (2003) for generating duplications in *S. enterica.*

Inversions

Making a defined inversion using recombineering is most easily achieved with a two-step process. In the first step, the region to be inverted is deleted, perhaps while inserting a selectable/counter-selectable cassette. In the second step, a PCR product of the region, containing the appropriate flanking homologies, is recombineered and replaces this counter-selectable cassette (or deleted region). The final product must be sequenced as PCR can create mutations. A similar approach was used for inverting the *gal* operon (Ellis *et al.*, 2001).

Annealing Oligos In Vivo

Two or more overlapping oligos can be simultaneously electroporated into Red-expressing cells. These oligos have two parts, an end with homol-ogy to the target sequence and an end complementary to the other oligo (Yu *et al.*, 2003). The oligos anneal *in vivo,* perhaps with the help of Beta, which would also protect them from degradation. The oligos must overlap by six or more bases to anneal and longer overlaps increase efficiency. If the annealed oligos have 5' single-stranded overhangs (the target homol-ogy), they recombine efficiently. Using this technique, multiple overlap-ping oligos can be used to construct longer DNA substrates (Yu *et al.,* 2003). This reaction is very similar to *in vitro* PCR assembly (Stemmer *et al.*, 1995) but occurs *in vivo.*

Gene-Specific Random Mutagenesis Using Recombineering

Recently, a useful protocol that includes recombineering to generate random, site-directed (a specific gene, for example) mutations has been published (De Lay and Cronan, 2006). Briefly, a mutagenized PCR product of your gene of interest is made (product 1). A PCR product of a nearby gene containing a selectable marker is also made (product 2). The two PCR products overlap by ~20 bases, are gel purified, mixed together, and overlapping extension PCR is performed (Ho *et al.*, 1989). Finally, the overlapping extension PCR product is used as a substrate for recombineering, inserting both the mutagenized fragment and the selectable marker into the chromosome. Mutations in your gene are then screened for. Such a targeted mutagenesis should be useful for many genes. De Lay and Cronan (2006) developed this technique to isolate temperature-sensitive mutations in an essential gene.

We imagine that gene-specific random mutagenesis can be done without a selectable marker. The gene can again be amplified by mutagenic PCR and used directly for recombineering in a mismatch-repair mutant host, thereby ensuring very high levels of recombination and relatively easy screening for mutant phenotypes.

Targeting Recombineering to Plasmids: Modifications to the Standard Protocol

Although we have emphasized modifying genes on the chromosome, the techniques discussed thus far can be used to modify plasmids as well. In addition, direct *in vivo* cloning can be accomplished with recombineering.

Recombineering on Plasmids

Recombineering targeted to a pBR322-type plasmid has been characterized, and frequencies similar to those obtained when targeting the *E. coli* chromosome are observed for both ds- and ss-DNA recombination. These results will be detailed in Thomason *et al.* (submitted), but the key findings are summarized here. It is critical to start with a pure monomer species plasmid for recombineering. Optimally, the plasmid should be introduced into *recA* mutant cells expressing the Red system by co-electroporation rather than targeting a resident plasmid. A low-plasmid DNA concentration should be used; 10 ng is usually sufficient for maximal transformation efficiency. After recombinant colonies are identified, they should be purified, under selective conditions if possible, before they are used to inoculate cultures from which to isolate candidate modified plasmid DNA. This DNA should be introduced into a *recA* mutant standard cloning strain at a low DNA concentration, once again selecting or screening for the desired modification.

Circular plasmid multimers arise when targeting plasmids. One source of these circular multimers is recombination catalyzed by the host RecA protein; these can be eliminated through the use of a *recA* mutant host for the recombineering. Another source of circular multimers is Red recombination acting on both double- and single-stranded linear substrate DNA; these multimers cannot be eliminated. In a *recA* mutant recombineering host, circular multimers are rarely, if ever, found among nonrecombinants. Co-electroporation of the plasmid with the modifying DNA minimizes, but does not eliminate, the formation of these plasmid multimers. It is important to screen recombinant plasmids by gel electrophoresis to determine their multimeric state. Multimeric recombinant plasmid products that have been converted on only one copy of the region to be altered have been observed. If a recombinant plasmid has multimerized, the DNA can be digested, re-ligated under dilute conditions, and then introduced into a *recA* mutant host lacking the Red system in order to obtain a recombinant monomer clone. It has been reported (Cohen and Clark, 1986) that extended expression of the Gam protein can give rise to linear plasmid multimers, but the circular multimers we have observed depend only on Beta expression and the presence of linear substrate DNA during recombination.

Gap Repair of Plasmids: In Vivo Cloning

Recombineering using a gapped plasmid with homology to the target can be used to clone genes or regions from the chromosome or other replicons (e.g., BACs). A gapped plasmid is a linear DNA fragment containing a plasmid origin. Gap repair of this linear plasmid is useful to retrieve a mutated gene for sequencing, allow expression of a gene under the control of a chosen promoter, or to create a gene fusion to a tag or reporter (Fig. 3). A gapped linear plasmid can also recombine with a co-electroporated linear fragment (Court *et al.*, 2002).

Oligos for PCR amplification of a gapped plasmid are designed as outlined in "preparing linear dsDNA." In this case, however, the 3' ends prime synthesis of a plasmid origin, and the 5' ends have homology flanking the target sequence to be cloned. Two methods can be used that differ in the location of the drug-resistance cassette, which can be either on the gapped plasmid itself or linked to the target sequence. If the target contains a linked drug-resistance cassette, the gapped plasmid need only contain a plasmid origin and homologies to the target (Datta *et al.*, 2006). The region to be cloned can be either co-transformed with the linear origin fragment or be already present in the cell. After recombineering, the electroporation mix is diluted in 10 ml of LB and incubated overnight nonselectively. Plasmid DNA is isolated and transformed into a standard *recA* mutant cloning strain using a low concentration of DNA to ensure that only one plasmid enters the cell. Select for the marker retrieved onto the origin

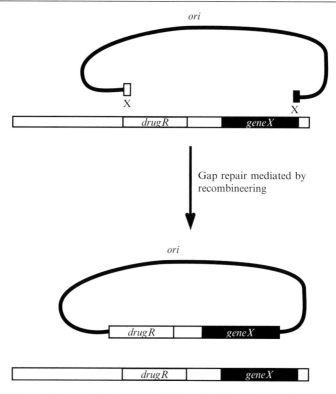

Fɪɢ. 3. Cloning by retrieval onto a gapped plasmid with recombineering. A linear DNA fragment containing a plasmid origin and homologies (∼50 bp on each end) to a region of interest can be used to clone sequences from the chromosome, other plasmids, BACs, or even a co-electroporated linear DNA fragment. In this illustration, a drug resistance is linked to the gene of interest, and thus the gapped linear plasmid need not contain a selectable marker. As the chromosome will still contain this drug resistance, plasmid DNA must be isolated and screened to find the desired recombinants as described in the text.

vector and confirm candidate recombinants by PCR (Thomason *et al.*, 2005a).

If there is no drug resistance linked to the target sequence, then the drug cassette must be on the gapped plasmid. The linear DNA "vector" containing a plasmid origin, a drug-resistance cassette, and ending in homologies to the target sequence is transformed into a cell that has been induced for the Red system. The target can be either co-transformed or already resident in the Red-producing strain and selection is for the drug resistance on the gapped plasmid. After purifying drug-resistant candidates, the recombinant plasmids must be checked since false positives can be caused by nonhomologous end joining of the linear vector. Repeats longer than 5 bp near the ends enhance nonhomologous end joining (Zhang *et al.*, 2000), which can be minimized by careful primer design.

Gap repair is less efficient than ssDNA or gene replacement recombineering; typically we see a few hundred recombinants per 10^8 viable cells. Because of this low frequency, it is important to eliminate false positives (see "Preparing Linear dsDNA" section). The small effort involved makes gap-repair cloning techniques very appealing as compared to traditional cloning methods. An important advantage is that DNA retrieved by gap repair from the chromosome is not subject to PCR-generated mutations.

Replacing Plasmid Origins

During genetic studies, one often encounters the problem that a plasmid is incompatible for use with another plasmid (or the chromosome) because it has the same drug resistance or origin of replication. Recombineering can be used to change the drug resistance of one of the plasmids. It can also be used to exchange one plasmid origin for another, thereby making one plasmid compatible with the other. Changing the origin can also be used to alter the copy number and/or extend the host range (Datta *et al.*, 2006).

Many clones are found in pBR322-based plasmids. Since the pBR322 origin does not replicate in a *polA* mutant strain (Kingsbury and Helinski, 1970), the origin of these plasmids can be selected against and replaced with other origins. A linear DNA fragment containing a new origin and any necessary replication functions (e.g., pSC101 or the RK2 origin) with homologies flanking the pBR322 origin can be electroporated into a strain containing the Red functions and the pBR322-based plasmid. After recombineering, the culture is grown nonselectively overnight in 10 ml of LB, plasmid DNA is prepared, and then used to transform a *polA* mutant strain with selection for the plasmid drug marker. Only plasmids that have acquired the new origin will be able to replicate in the *polA* mutant strain (Datta *et al.*, 2006). We note that origin replacement is mechanistically the same as retrieval by gap repair.

Targeting Recombineering to Phage: Modifications to the Standard Protocol

Like other replicons, the phage λ chromosome can also be modified by recombineering. The Red proteins can be supplied by a prophage on the chromosome (Court *et al.*, 2003; Oppenheim *et al.*, 2004), by a defective prophage on a plasmid (Datta *et al.*, 2006) or by the infecting phage itself (Oppenheim and Costantino, unpublished results). The "standard recombineering protocol" has been modified (Oppenheim *et al.*, 2004). Cells containing a defective prophage are grown to mid-log at 32° and then harvested by centrifugation at $4600 \times g$ for 7 min at $4°$ before suspending them in 1 ml of TMG buffer (see "Media" section). The λ phage to be modified are added at a multiplicity of infection of one to three phage per cell and are adsorbed at room temperature for 15 min. The infected cells are added to 5 ml of LB prewarmed to $42°$, which will induce production of the Red proteins. Cultures are shaken in baffled flasks at 200 rpm for 15 min. After 15 min, the cultures are chilled on ice and processed as described in the standard recombineering protocol. The cells are electroporated with either a PCR product or oligo, diluted into 5 ml of $39°$ LB, and allowed to incubate at $39°$ with shaking to finish the lambda lytic cycle (60 to 90 min). As a negative control, include an electroporation without PCR or oligo. The lysates are diluted and titered on appropriate bacteria to obtain single plaques. The desired mutation can be selected or screened for (Oppenheim et al., 2004).

If recombineering is done with an intact (cI857) prophage, then induction at $42°$ should only be for 4 to 5 min to prevent cell killing. The shorter time minimizes expression of the prophage DNA replication genes, which are toxic to the host when expressed for longer periods (Court et al., 2003). The rest of the protocol is as outlined.

Prophage-Containing Recombineering Plasmids

Recombineering has already proven very useful for bacterial genetics in E. coli, pathogenic E. coli (Murphy and Campellone, 2003), and S. enterica (Bunny et al., 2002; Uzzau et al., 2001). This technology has also been used to modify plasmids or BACs in E. coli before moving the altered constructs to other organisms such as mice (Lee et al., 2001; Warming et al., 2005), and Aspergillus nidulans (Chaveroche et al., 2000). Pioneering studies have been done in Yersinia pseudotuberculosis (Derbise et al., 2003), and will undoubtedly be tried in other prokaryotes and perhaps eukaryotes soon.

Recently, we have made a series of plasmids that should aid recombineering in E. coli and certain other gram-negative bacteria (Datta et al., 2006). These plasmids contain a defective prophage in which the pL promoter has been directly fused to the Red genes, thereby removing some of the normal regulatory elements (Fig. 1B). The pL promoter and Red expression on these plasmid vectors are still tightly regulated by the temperature-sensitive repressor, CI857. These vectors are available with different plasmid origins of replication and drug-resistant markers as described in Table II. Another vector, mini-λ, was developed to move the defective prophage system between E. coli strains (Court et al., 2003). However, the plasmid vectors just described are more efficient for this purpose and still maintain tight control of Red gene expression.

Strains and Plasmids

Many bacterial strains and plasmid vectors that are useful for using λ Red recombineering have been constructed. Table IV lists several recombineering strains and their genotypes. Table II describes the key attributes of several recombineering plasmids that are currently available.

Media

The growth media for the various protocols, in quantities per liter, follow. As indicated in Table IV, many recombineering strains are biotin auxotrophs, and biotin must be added to a final concentration of 0.0001% (w/v) to all minimal media.

Luria Broth (LB)

10 g Bacto-typtone (Difco)
5 g yeast extract (Difco)
5 g NaCl (not 10 g, as used by many)

Note: Add 15 g Bacto-agar (Difco) for plates.

L + Sucrose (No NaCl) Plates

L plates are supplemented with 6% (w/v) sucrose for selecting against *sacB*. NaCl should be omitted from this medium (Blomfield *et al.*, 1991).

M63 Minimal Glycerol + Sucrose Plates

3 g KH_2PO_4
7 g K_2HPO_4
2 g $(NH_4)_2SO_4$
0.5 ml $FeSO_4$ (1 mg/ml solution)
1 ml $1M$ $MgSO_4$
10 ml 20% glycerol
5% (w/v) sucrose
5 ml 0.2 mg/ml (0.02%) D-biotin (Sigma)
1 ml 1% thiamine (vitamin B_1)
15 g Bacto-Agar

M63-DOG (for Selecting GalK Mutants)

3 g KH_2PO_4
7 g K_2HPO_4
2 g $(NH_4)_2SO_4$

TABLE IV

Useful Recombineering Strains

Strain	Genotype	Special purpose	References
DY329	**W3110** Δ*lacU169 nadA::Tn10 gal490 pgl*Δ*8* [λ *cl857* Δ*(cro bioA)*]	Useful for moving prophage by P1 transduction using linked *Tn10*	(Yu *et al.*, 2000)
DY330	**W3110** Δ*lacU169 gal490 pgl*Δ*8* [λ *cl857* Δ*(cro-bioA)*]		(Yu *et al.*, 2000)
DY331	**W3110** Δ*lacU169* Δ*(srlA-recA)301::Tn10 gal490 pgl*Δ*8* [λ *cl857* Δ*(cro-bioA)*]		(Yu *et al.*, 2000)
DY378	**W3110** [λ *cl857* Δ*(cro-bioA)*]		(Yu *et al.*, 2000)
DY380	**DH10B** *mcrA* Δ*(mrr-hsdRMS-mcrBC)* φ*80dlacZ*Δ*M15* Δ*lacX74 deoR recA1 endA1 araD139* Δ*(ara, leu)7697 galU gal490 pgl*Δ*8 rpsL nupG* [λ *cl857ind1* Δ*(cro-bioA)<>tet*]	Useful for BAC transformation and manipulations	(Lee *et al.*, 2001)
HME5	**W3110** Δ*lacU169* [λ *cl857* Δ*(cro-bioA)*]		(Ellis *et al.*, 2001)
HME6	**W3110** *galK*_{tyr145UAG} Δ*lacU169* [λ *cl857* Δ*(cro-bioA)*]	Assay system for oligo recombineering.	(Ellis *et al.*, 2001)
HME43	**W3110** *galK*_{tyr145UAG} Δ*lacU169* [λ *cl857* Δ*(exo-int)<>cat* Δ*<>{gam-N}*]	Strain makes only Red Beta	(Ellis *et al.*, 2001)
HME51	**W3110** *galK*_{tyr145UAG} Δ*lacU169* [λ *cl857* Δ*(exo-int)<>cat* Δ*<>{gam-N}*] Δ*(srlA-recA)301::Tn10*	Defective for mismatch repair; therefore, high-level oligo recombineering	N. Costantino, personal communication
HME63	**W3110** *galK*_{tyr145UAG} Δ*lacU169 mutS>amp* [λ *cl857* Δ*(cro-bioA)*]	Defective for mismatch repair	(Costantino and Court, 2003)
HME68	**W3110** *galK*_{tyr145UAG} Δ*lacU169* [λ *cl857* Δ*(cro-bioA)*] *mutS<>cat*		N. Costantino, personal communication
HME70	**W3110** *galK*_{tyr145UAG} Δ*lacU169* [λ *cl857* Δ*(cro-bioA)*] *mutS<>cat* Δ*(srlA-recA)301::Tn10*	Oligo recombineering with plasmids	(Thomason *et al.*, submitted)
HME71	**W3110** *galK*_{tyr145UAG} Δ*lacU169* [λ *cl857* Δ*(cro-bioA)*] Δ*(srlA-recA)301::Tn10*	Oligo recombineering with plasmids	N. Costantino, personal communication
SIMD3	**W3110** [λ *cl857* Δ*rex<>cat* Δ*(N-kil)* Δ*(cro-bioA)*]	Contains N-independent minimal prophage (Fig. 1B)	(Datta *et al.*, 2006)
SIMD4	**W3110** [λ *cl857* Δ*rex<>amp* Δ*(N-kil)* Δ*(cro-bioA)*]	Contains N-independent minimal prophage (Fig. B)	(Datta *et al.*, 2006)
SW102	**DH10B** *mcrA* Δ*(mrr-hsdRMS-mcrBC)* φ*80dlacZ*Δ*M15* Δ*lacX74 deoR recA1 endA1 araD139* Δ*(ara, leu)7697* Δ*galK pgl*Δ*8 rpsL nupG* [λ *cl857ind1* Δ*(cro-bioA)<>tet*]	Use for *galK* selection/counter-selection	(Warming *et al.*, 2005)

0.5 ml $FeSO_4$ (1 mg/ml solution)
1 ml 1 M $MgSO_4$
10 ml 20% glycerol
5 ml 0.2 mg/ml (0.02%) D-biotin (Sigma)
1 ml 1% thiamine (vitamin B_1)
5 ml 40% 2-deoxy-galactose (DOG) (Ferro Pfanstiehl)
15 g Bacto-agar

TMG Buffer

10 mM Tris base
10 mM $MgSO_4$
0.01% gelatin

Antibiotics

When antibiotics are added to select for single copy markers (i.e., on the chromosome), they are used at lower concentrations than for plasmid selection. Using a too-high drug concentration will reduce the number or even prevent detection of recombinants. The following is for single copy use: ampicillin, 30 μg/ml; chloramphenicol, 10 μg/ml; kanamycin, 30 μg/ml; tetracycline, 12.5 μg/ml; and spectinomycin, 30 to 50 μg/ml. These concentrations have been used in *E. coli* and *S. enterica*, but the proper concentrations in other bacteria must be determined.

Concluding Remarks

Recombineering has made complex genetic manipulations possible. Large DNA molecules such as BACs and the chromosome can be directly modified. In contrast to site-specific recombination systems that leave a *loxP* or *frt* site at the modified region, recombineering does not necessarily leave "scars" behind. Although recombineering has been primarily developed in *E. coli*, it is starting to be used in other bacteria and soon perhaps even in eukaryotes. New advances in the understanding of the mechanisms as well as new ways to use recombineering are rapidly being developed. See http://RedRecombineering.ncifcrf.gov/ and http://recombineering.ncifcrf.gov/ to download protocols as well as to check for updates of techniques, and to request strains or plasmids.

Acknowledgments

This research was supported by the Intramural Research Program of the National Institutes of Health (NIH), National Cancer Institute, Center for Cancer Research, and in part by a Trans NIH/Food and Drug Administration Intramural Biodefense Program Grant from the National Institute of Allergy and Infectious Diseases to D. L. Court.

References

Ahmed, A., and Podemski, L. (1995). The revised nucleotide sequence of *Tn5*. *Gene* **154,** 129–130.

Alper, M. D., and Ames, B. N. (1975). Positive selection of mutants with deletions of the *gal-chl* region of the *Salmonella* chromosome as a screening procedure for mutagens that cause deletions. *J. Bacteriol.* **121,** 259–266.

Arber, W., Enquist, L., Hohn, B., Murray, N., and Murray, K. (1983). Experimental methods for use with lambda. *In* "Lambda II" (R. W. Hendrix, J. W. Roberts, F. W. Stahl, and R. A. Weisberg, eds.). Cold Spring Harbor Laboratory Press, Cold Spring Harbor, NY.

Blomfield, I. C., Vaughn, V., Rest, R. F., and Eisenstein, B. I. (1991). Allelic exchange in *Escherichia coli* using the *Bacillus subtilis sacB* gene and a temperature-sensitive pSC101 replicon. *Mol. Microbiol.* **5,** 1447–1457.

Bunny, K., Liu, J., and Roth, J. (2002). Phenotypes of *lexA* mutations in *Salmonella enterica*: Evidence for a lethal *lexA* null phenotype due to the Fels-2 prophage. *J. Bacteriol.* **184,** 6235–6249.

Carter, D. M., and Radding, C. M. (1971). The role of exonuclease and β protein of phage λ in genetic recombination. II. Substrate specificity and the mode of action of lambda exonuclease. *J. Biol. Chem.* **246,** 2502–2512.

Cassuto, E., and Radding, C. M. (1971). Mechanism for the action of λ exonuclease in genetic recombination. *Nat. New Biol.* **229,** 13–16.

Cassuto, E., Lash, T., Sriprakash, K. S., and Radding, C. M. (1971). Role of exonuclease and β protein of phage λ in genetic recombination. V. Recombination of λ DNA *in vitro*. *Proc. Natl. Acad. Sci. USA* **68,** 1639–1643.

Cha, R. S., Zarbl, H., Keohavong, P., and Thilly, W. G. (1992). Mismatch amplification mutation assay (MAMA): Application to the c-H-ras gene. *PCR Methods Appl.* **2,** 14–20.

Chalker, A. F., Leach, D. R., and Lloyd, R. G. (1988). *Escherichia coli sbcC* mutants permit stable propagation of DNA replicons containing a long palindrome. *Gene* **71,** 201–205.

Chaveroche, M. K., Ghigo, J. M., and d'Enfert, C. (2000). A rapid method for efficient gene replacement in the filamentous fungus *Aspergillus nidulans*. *Nucleic Acids Res.* **28,** E97.

Cohen, A., and Clark, A. J. (1986). Synthesis of linear plasmid multimers in *Escherichia coli* K-12. *J. Bacteriol.* **167,** 327–335.

Copeland, N. G., Jenkins, N. A., and Court, D. L. (2001). Recombineering: A powerful new tool for mouse functional genomics. *Nat. Rev. Genet.* **2,** 769–779.

Costantino, N., and Court, D. L. (2003). Enhanced levels of λ Red-mediated recombinants in mismatch repair mutants. *Proc. Natl. Acad. Sci. USA* **100,** 15748–15753.

Court, D. L., Sawitzke, J. A., and Thomason, L. C. (2002). Genetic engineering using homologous recombination. *Ann. Rev. Genet.* **36,** 361–388.

Court, D. L., Swaminathan, S., Yu, S., Wilson, H., Baker, T., Bubunenko, M., Sawitzke, J., and Sharan, S. K. (2003). Mini-λ: A tractable system for chromosome and BAC engineering. *Gene* **315,** 63–69.

Datsenko, K. A., and Wanner, B. L. (2000). One-step inactivation of chromosomal genes in *Escherichia coli* K-12 using PCR products. *Proc. Natl. Acad. Sci. USA* **97,** 6640–6645.

Datta, S., Costantino, N., and Court, D. L. (2006). A set of recombineering plasmids for gram-negative bacteria. *Gene.* **379,** 109–115.

De Lay, N. R., and Cronan, J. E. (2006). Gene-specific random mutagenesis of *Escherichia coli in vivo*: Isolation of temperature-sensitive mutations in the acyl carrier protein of fatty acid synthesis. *J. Bacteriol.* **188,** 287–296.

Derbise, A., Lesic, B., Dacheux, D., Ghigo, J. M., and Carniel, E. (2003). A rapid and simple method for inactivating chromosomal genes in *Yersinia*. *FEMS Immunol. Med. Microbiol.* **38,** 113–116.

Dohet, C., Wagner, R., and Radman, M. (1986). Methyl-directed repair of frameshift mutations in heteroduplex DNA. *Proc. Natl. Acad. Sci. USA* **83,** 3395–3397.

Ellis, H. M., Yu, D., DiTizio, T., and Court, D. L. (2001). High efficiency mutagenesis, repair, and engineering of chromosomal DNA using single-stranded oligonucleotides. *Proc. Natl. Acad. Sci. USA* **98,** 6742–6746.

Friedman, S. A., and Hays, J. B. (1986). Selective inhibition of *Escherichia coli recBC* activities by plasmid-encoded GamS function of phage λ. *Gene* **43,** 255–263.

Gay, P., Le Coq, D., Steinmetz, M., Berkelman, T., and Kado, C. I. (1985). Positive selection procedure for entrapment of insertion sequence elements in gram-negative bacteria. *J. Bacteriol.* **164,** 918–921.

Gibson, F. P., Leach, D. R. F., and Lloyd, R. G. (1992). Identification of *sbcD* mutations as cosuppressors of *recBC* that allow propagation of DNA palindromes in *Escherichia coli* K-12. *J. Bacteriol.* **174,** 1222–1228.

Haack, K. R., and Roth, J. R. (1995). Recombination between chromosomal IS200 elements supports frequent duplication formation in *Salmonella typhimurium*. *Genetics* **141,** 1245–1252.

Haber, L. T., and Walker, G. C. (1991). Altering the conserved nucleotide binding motif in the *Salmonella typhimurium* MutS mismatch repair protein affects both its ATPase and mismatch binding activities. *EMBO J.* **10,** 2707–2715.

Hillen, W., and Schollmeier, K. (1983). Nucleotide sequence of the *Tn10* encoded tetracycline resistance gene. *Nucleic Acids Res.* **11,** 525–539.

Ho, S. N., Hunt, H. D., Horton, R. M., Pullen, J. K., and Pease, L. R. (1989). Site-directed mutagenesis by overlap extension using the polymerase chain reaction. *Gene* **77,** 51–59.

Karakousis, G., Ye, N., Li, Z., Chiu, S. K., Reddy, G., and Radding, C. M. (1998). The β protein of phage λ binds preferentially to an intermediate in DNA renaturation. *J. Mol. Biol.* **276,** 721–731.

Karu, A. E., Sakaki, Y., Echols, H., and Linn, S. (1975). The γ protein specified by bacteriophage λ. Structure and inhibitory activity for the RecBC enzyme of *Escherichia coli*. *J. Biol. Chem.* **250,** 7377–7387.

Kingsbury, D. T., and Helinski, D. R. (1970). DNA polymerase as a requirement for the maintenance of the bacterial plasmid colicinogenic factor E1. *Biochem. Biophys. Res. Commun.* **41,** 1538–1544.

Kovach, M. E., Phillips, R. W., Elzer, P. H., Roop, R. M., 2nd, and Peterson, K. M. (1994). pBBR1MCS: A broad-host-range cloning vector. *BioTechniques* **16,** 800–802.

Kulkarni, S. K., and Stahl, F. W. (1989). Interaction between the *sbcC* gene of *Escherichia coli* and the *gam* gene of phage λ. *Genetics* **123,** 249–253.

Lee, E. C., Yu, D., Martinez de Velasco, J., Tessarollo, L., Swing, D. A., Court, D. L., Jenkins, N. A., and Copeland, N. G. (2001). A highly efficient *Escherichia coli*-based chromosome engineering system adapted for recombinogenic targeting and subcloning of BAC DNA. *Genomics* **73,** 56–65.

Li, X. T., Costantino, N., Lu, L. Y., Liu, D. P., Watt, R. M., Cheah, K. S., Court, D. L., and Huang, J. D. (2003). Identification of factors influencing strand bias in oligonucleotide-mediated recombination in *Escherichia coli*. *Nucleic Acids Res.* **31,** 6674–6687.

Li, Z., Karakousis, G., Chiu, S. K., Reddy, G., and Radding, C. M. (1998). The β protein of phage λ promotes strand exchange. *J. Mol. Biol.* **276,** 733–744.

Muniyappa, K., and Radding, C. M. (1986). The homologous recombination system of phage λ. Pairing activities of β protein. *J. Biol. Chem.* **261,** 7472–7478.

Murphy, K. C. (1991). λ Gam protein inhibits the helicase and chi-stimulated recombination activities of *Escherichia coli* RecBCD enzyme. *J. Bacteriol.* **173,** 5808–5821.

Murphy, K. C., and Campellone, K. G. (2003). Lambda Red-mediated recombinogenic engineering of enterohemorrhagic and enteropathogenic *E. coli.*. *BMC Mol. Biol.* **4,** 11.

Muyrers, J. P., Zhang, Y., Testa, G., and Stewart, A. F. (1999). Rapid modification of bacterial artificial chromosomes by ET-recombination. *Nucleic Acids Res.* **27,** 1555–1557.

Muyrers, J. P., Zhang, Y., Buchholz, F., and Stewart, A. F. (2000). RecE/RecT and Redα/Redβ initiate double-stranded break repair by specifically interacting with their respective partners. *Genes Dev.* **14,** 1971–1982.

Mythili, E., Kumar, K. A., and Muniyappa, K. (1996). Characterization of the DNA-binding domain of β protein, A component of phage λ Red-pathway, by UV catalyzed cross-linking. *Gene* **182,** 81–87.

Oppenheim, A. B., Rattray, A. J., Bubunenko, M., Thomason, L. C., and Court, D. L. (2004). *In vivo* recombineering of bacteriophage λ by PCR fragments and single-strand oligonucleotides. *Virology* **319,** 185–189.

Poteete, A. R., and Fenton, A. C. (1993). Efficient double-strand break-stimulated recombination promoted by the general recombination systems of phages λ and P22. *Genetics* **134,** 1013–1021.

Poteete, A. R. (2001). What makes the bacteriophage λ Red system useful for genetic engineering: Molecular mechanism and biological function. *FEMS Microbiol. Lett.* **201,** 9–14.

Radding, C. M., Rosenzweig, J., Richards, F., and Cassuto, E. (1971). Separation and characterization of exonuclease, β protein, and a complex of both. *J. Biol. Chem.* **246,** 2510–2512.

Sambrook, J., and Russell, D. W. (2001). "Molecular Cloning: A Laboratory Manual." Cold Spring Harbor Laboratory Press, Cold Spring Harbor, NY.

Sergueev, K., Yu, D., Austin, S., and Court, D. (2001). Cell toxicity caused by products of the *pL* operon of bacteriophage lambda. *Gene* **272,** 227–235.

Sergueev, K., Court, D., Reaves, L., and Austin, S. (2002). *E. coli* cell-cycle regulation by bacteriophage lambda. *J. Mol. Biol.* **324,** 297–307.

Shen, P., and Huang, H. V. (1986). Homologous recombination in *Escherichia coli*: Dependence on substrate length and homology. *Genetics* **112,** 441–457.

Silberstein, Z., and Cohen, A. (1987). Synthesis of linear multimers of OriC and pBR322 derivatives in *Escherichia coli* K-12: Role of recombination and replication functions. *J. Bacteriol.* **169,** 3131–3137.

Silberstein, Z., Maor, S., Berger, I., and Cohen, A. (1990). λ Red-mediated synthesis of plasmid linear multimers in *Escherichia coli* K12. *Mol. Gen. Genet.* **223,** 496–507.

Slechta, E. S., Bunny, K. L., Kugelberg, E., Kofoid, E., Andersson, D. I., and Roth, J. R. (2003). Adaptive mutation: General mutagenesis is not a programmed response to stress but results from rare coamplification of *dinB* with *lac*. *Proc. Natl. Acad. Sci. USA* **100,** 12847–12852.

Stahl, F. W., Kobayashi, I., and Stahl, M. M. (1985). In phage λ, *cos* is a recombinator in the Red pathway. *J. Mol. Biol.* **181,** 199–209.

Stemmer, W. P., Crameri, A., Ha, K. D., Brennan, T. M., and Heyneker, H. L. (1995). Single-step assembly of a gene and entire plasmid from large numbers of oligodeoxyribonucleotides. *Gene* **147,** 49–53.

Su, S. S., Lahue, R. S., Au, K. G., and Modrich, P. (1988). Mispair specificity of methyl-directed DNA mismatch correction *in vitro*. *J. Biol. Chem.* **263,** 6829–6835.

Swaminathan, S., Ellis, H. M., Waters, L. S., Yu, D., Lee, E.-C., Court, D. L., and Sharan, S. K. (2001). Rapid engineering of bacterial artificial chromosomes using oligonucleotides. *Genesis* **29,** 14–21.

Thaler, D. S., Stahl, M. M., and Stahl, F. W. (1987a). Tests of the double-strand-break repair model for Red-mediated recombination of phage λ and plasmid λ *dv*. *Genetics* **116,** 501–511.

Thaler, D. S., Stahl, M. M., and Stahl, F. W. (1987b). Double-chain–cut sites are recombination hotspots in the Red pathway of phage λ. *J. Mol. Biol.* **195,** 75–87.

Thomason, L. C., Bubunenko, M., Costantino, N., Wilson, H., Oppenheim, A. B., Datta, S., and Court, D. L. (2005a). Recombineering: Genetic engineering in bacteria using homologous recombination. *In* "Current Protocols in Molecular Biology" (F. M. Ausubel and R. Brent, eds.). John Wiley and Sons, Hoboken, NJ.

Thomason, L. C., Myers, R. S., Oppenheim, A., Costantino, N., Sawitzke, J. A., Datta, S., Bubunenko, M., and Court, D. L. (2005b). Recombineering in prokaryotes. *In* "Phages: Their Role in Bacterial Pathogenesis and Biotechnology" (M. K. Waldor, D. I. Friedman, and S. L. Adhya, eds.). ASM Press, Washington, DC.

Uzzau, S., Figueroa-Bossi, N., Rubino, S., and Bossi, L. (2001). Epitope tagging of chromosomal genes in *Salmonella. Proc. Natl. Acad. Sci. USA* **98,** 15264–15269.

Vellani, T. S., and Myers, R. S. (2003). Bacteriophage SPP1 Chu is an alkaline exonuclease in the SynExo family of viral two-component recombinases. *J. Bacteriol.* **185,** 2465–2474.

Warming, S., Costantino, N., Court, D. L., Jenkins, N. A., and Copeland, N. G. (2005). Simple and highly efficient BAC recombineering using *galK* selection. *Nucleic Acids Res.* **33,** e36.

Watt, V. M., Ingles, C. J., Urdea, M. S., and Rutter, W. J. (1985). Homology requirements for recombination in *Escherichia coli. Proc. Natl. Acad. Sci. USA* **82,** 4768–4772.

Wong, Q. N., Ng, V. C., Lin, M. C., Kung, H. F., Chan, D., and Huang, J. D. (2005). Efficient and seamless DNA recombineering using a thymidylate synthase A selection system in *Escherichia coli. Nucleic Acids Res.* **33,** e59.

Yang, Y., and Sharan, S. K. (2003). A simple two-step, "hit and fix" method to generate subtle mutations in BACs using short denatured PCR fragments. *Nucleic Acids Res.* **31,** e80.

Yu, D., Ellis, H. M., Lee, E. C., Jenkins, N. A., Copeland, N. G., and Court, D. L. (2000). An efficient recombination system for chromosome engineering in *Escherichia coli. Proc. Natl. Acad. Sci. USA* **97,** 5978–5983.

Yu, D., Sawitzke, J. A., Ellis, H., and Court, D. L. (2003). Recombineering with overlapping single-stranded DNA oligonucleotides: Testing a recombination intermediate. *Proc. Natl. Acad. Sci. USA* **100,** 7207–7212.

Zhang, Y., Buchholz, F., Muyrers, J. P., and Stewart, A. F. (1998). A new logic for DNA engineering using recombination in *Escherichia coli. Nature Genetics* **20,** 123–128.

Zhang, Y., Muyrers, J. P., Testa, G., and Stewart, A. F. (2000). DNA cloning by homologous recombination in *Escherichia coli. Nat. Biotechnol.* **18,** 1314–1317.

[16] λ-Red Genetic Engineering in *Salmonella enterica* serovar Typhimurium

By Joyce E. Karlinsey

Abstract

The use of the recombination system from bacteriophage lambda, λ-Red, allows for PCR-generated fragments to be targeted to specific chromosomal locations in sequenced genomes. A minimal region of homology of 30 to 50 bases flanking the fragment to be inserted is all that is required for targeted mutagenesis. Procedures for creating specific insertions, deletions, and site-directed changes are described.

METHODS IN ENZYMOLOGY, VOL. 421   0076-6879/07 $35.00
DOI: 10.1016/S0076-6879(06)21016-4

Introduction

Phage-based homologous recombination systems are emerging as a versatile and alternative method for chromosomal genetic engineering (Copeland *et al.*, 2001; Court *et al.*, 2002; Muyrers *et al.*, 2001; Poteete, 2001). This method utilizes phage recombination proteins that require only short regions of homology at the 5' ends of linear DNA fragments to mediate homologous recombination in the cell. This streamlines chromosomal genetic engineering by circumventing time-consuming *in vitro* DNA cloning. Site-specific chromosomal gene knockouts are easily procured by introducing selectable markers, such as antibiotic-resistant genes, with short homologous flanking DNA sequences into cells expressing the phage recombination proteins and selection on the appropriate medium (Datsenko and Wanner, 2000). In addition, the construction of in-frame gene deletions, site-specific mutations and gene fusions in the genome by phage-mediated recombination are performed *in vivo*, which eliminates the need for several DNA cloning steps (Datsenko and Wanner, 2000; Karlinsey and Hughes, 2006; Uzzau *et al.*, 2001). These systems have also been well adapted for manipulating large DNA fragments that are needed in the genetic engineering of bacterial plasmids, bacterial artificial chromosomes, and genomes (Lee *et al.*, 2001; Liu *et al.*, 2003; Poteete *et al.*, 2004; Zhang *et al.*, 1998). Further, phage-based homologous recombination have been successfully utilized in pathogenic organisms such as *Escherichia, Salmonella, Shigella*, and *Yersinia* species (Derbise *et al.*, 2003; Hu *et al.*, 2003; Murphy and Campellone, 2003; Uzzau *et al.*, 2001).

The λ-Red from phage λ and RecET from *Escherichia coli* Rac prophage are two phage-based homologous recombination systems that are commonly available. In both phage-based systems, recombination of linear double-stranded DNA (dsDNA) requires as little as 30 bps of homology (Yu *et al.*, 2000; Zhang *et al.*, 1998). The λ-Red genes (i.e., *gam, bet*, and *exo*) are the only genes needed to promote phage-dependent homologous recombination of linear dsDNA in cells (Yu *et al.*, 2000). Further, λ-Red–dependent recombination can occur in the absence of RecA (Stahl *et al.*, 1997). The expression of Gam inhibits the nuclease activities of RecBCD and presumably prevents degradation of linear dsDNA in the cell (Karu *et al.*, 1975; Murphy, 1991). Additionally, the RecBCD-dependent pathway for homologous recombination is reduced (Murphy, 1991). Exo is a 5'-3' dsDNA-dependent exonuclease that generates 3' ssDNA ends on linear dsDNA (Little, 1967). The 3' ssDNA ends are bound by Bet that catalyzes strand annealing and exchange (Li *et al.*, 1998). Recombinational frequencies as high as 1% per surviving electroporated cells of linear dsDNA have been reported (Murphy *et al.*, 2000). The λ-Red expression systems are available as prophages or on plasmids for a variety of applications (Datsenko and Wanner, 2000; Murphy and Campellone, 2003; Murphy *et al.*, 2000; Yu *et al.*, 2000).

The RacET-phage homologous system was found as another class of *recBC* suppressor, *scbA*, in *E. coli* K12 harboring the Rac prophage. The *scbA* mutation allowed the expression of Rac prophage genes *recE* and *recT* in *E. coli* (Barbour *et al.*, 1970). Rac phage genes *recE* and *recT* have homologous functions to the λ-Red genes *exo* and *bet*, and therefore represent another group of phage-encoded genes that allow homologous recombination of linear dsDNA in the cell (Kolodner *et al.*, 1994). Because the RacET systems lack λ *gam* function, recombination must be performed in *recBC* strains or by using plasmids expressing both RecET and λ Gam (Zhang *et al.*, 1998).

λ-Red Mediated Homologous Recombination in S. *typhimurium*

Our strategy for using λ-Red mediated homologous recombination for phage-mediated recombination in *Salmonella enterica* serovar Typhimurium (*S. typhimurium*) is shown in Fig. 1. The protocol utilizes the λ-Red phage recombination proteins, a selectable marker for allelic exchange and the ability to counter-select the selectable marker (Karlinsey and Hughes, 2006). The selectable marker employed is the *tetRA* element that encompasses the coding sequences from the *tetR* and *tetA* genes, and confers tetracycline resistance (Tc^R) from transposon Tn*10* (Fig. 2) (Way *et al.*, 1984). The *tetA* gene encodes a membrane-embedded antiporter protein TetA that renders the cell resistant to tetracycline by exporting tetracycline out of the cell (Yamaguchi *et al.*, 1990). The *tetR* gene encodes a repressor TetR that binds to operator sites and inhibits both *tetR* and *tetA* expression. When tetracycline is present, it binds to TetR and repression is relieved to allow transcription out from both genes (Way *et al.*, 1984).

To facilitate insertion of the PCR-amplified *tetRA* element, the *tetRA* sequence is flanked by 40 bps of homology to the targeted allele by PCR (Fig. 1). Primers are designed with 18 bases of homology to either the 3′ end of *tetR* or the 5′ end of *tetA* flanked by 40 bases of identity to the target site. The *tetRA* fragment with flanked DNA donor is electroporated into *S. typhimurium*—containing plasmid pKD46 (Datsenko and Wanner, 2000) expressing the λ-Red phage recombination proteins under an arabinose-inducible promoter (Fig. 1). Allelic replacement of the *tetRA* onto the chromosome is selected by screening for Tc^R colonies (Fig. 1). Plasmid pKD46 has a temperature-sensitive replicon and can be easily eliminated from the cell by incubation at 42° (Datsenko and Wanner, 2000). We have also constructed pJK611where the *Bacillus subtilis sacBR* genes from pRL250 (Cai and Wolk, 1990) were cloned into pKD46 (Karlinsey and

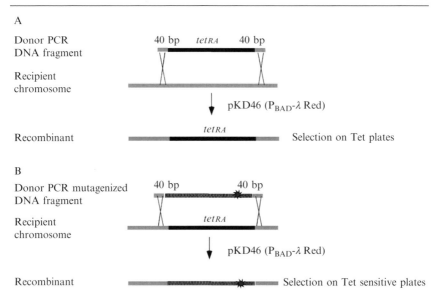

Fig. 1. General strategy for λ-Red genetic engineering in *S. enterica*. (A) A *tetRA* element is flanked by 40 bps of homology by PCR. λ-Red recombination proteins are expressed from plasmid pKD46 and facilitate recombination of the *tetRA* element into the chromosome. Recombinants are selected for on tetracycline medium. (B) The *tetRA* element is counter-selected by λ-Red recombination with a user-defined DNA fragment flanked by 40 bps of homology and selection on tetracycline-sensitive medium.

Hughes, personal communication). The plasmid is lost by growth on plates containing 5% sucrose, and is useful for phage-mediated recombination in temperature-sensitive strains where maintaining strains at the permissive temperature is essential.

To facilitate *in vivo* gene constructions of in-frame deletions, gene fusions, and site-directed mutagenesis, the *tetRA* element can be counter-selected by growth on tetracycline-sensitive (Tc^S) medium (Karlinsey and Hughes, 2006; Maloy and Nunn, 1981) (Fig. 1). To construct precise in-frame deletions, the gene is deleted and replaced with a *tetRA* element at deletion endpoints defined by the user. The *tetRA* element is subsequently replaced with an 80mer oligonucleotide that spans 40 bps up- and down-stream of the insertion site of the *tetRA* element using λ-Red recombination and selection on Tc^S plates. This basic strategy of counter-selecting the *tetRA* element with any user-defined DNA fragment allows for construction of deletions with precise endpoints and complete removal of the inserted material. This method allows the construction of strains with a multiplicity of chromosomal genetic alterations without leaving scars from previously inserted DNA fragments.

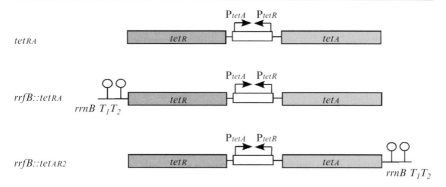

Fig. 2. The *tetRA* elements for λ-Red genetic engineering. The tetracycline-resistant regulon (*tetRA*) from transposon Tn*10* includes overlapping divergent promoters. The *tetRA* element includes the gene encoding for the tetracycline efflux pump, *tetA*, and its negative regulator, *tetR*. The *tetRA* element can be placed upstream of a given gene such that the gene is transcribed from the *tetA* or *tetR* promoter in an operon fusion. Two other forms of *tetRA* also have transcriptional terminators from the *rrnB* operon to prevent unwanted transcription into the *tetR* or *tetA* gene of the *tetAR* element. The *rrfB::tetRA* element has the *rrnB* T_1T_2 transcriptional terminators from *E. coli* downstream of *tetR* whereas the *rrfB*::tetAR2 element has the transcriptional terminators downstream *tetA*.

The *tetRA* element can also be used to create operon fusions that are solely regulated by the addition of tetracycline. The *tetR* and *tetA* genes share overlapping divergent promoters (Fig. 2). Further, upon tetracycline induction, these constructs allow directional read-out from either promoter of *tetA* or *tetR* (Fig. 2). The *tetA* promoter has been reported to be 7 to 11 times more active than the *tetR* promoter (Daniels and Bertrand, 1985); therefore, these constructs allow choices for promoter strengths for tetracycline-dependent expression. Two additional *tetRA* elements were constructed where the transcriptional terminator of *E. coli rrfB* (*rrnB* T_1T_2) was placed upstream of *tetA* and *tetR* to prevent potential transcriptional read-through from upstream promoters (Fig. 2) (Wozniak and Hughes, personal communication).

Procedures

Reagents

TAQ polymerase (preferably a proofreading TAQ)
10× PCR buffer
dNTP mix
Electrophoresis-grade agarose

TAE buffer (40 mM Tris-acetate/1 mM EDTA)

Luria broth (LB) (10 g Bacto-tryptone/5 g Bacto yeast extract/5 g NaCl per liter)

L-plates (LB + 1.5% Bacto agar)

Ampicillin (Ap)

Tetracycline (Tc)

L-arabinose (20% filter sterilize)

Sterile water

1-mm electroporation cuvetts

Primer Design for tetRA *Elements and for Allelic Replacement or Insertion*

1. Primer tetR, 5′-40-bps homology on sense strand + TTA AGA CCC ACT TTC ACA TT-3′.
2. Primer tetA, 5′-40-bps homology on nonsense strand + CTA AGC ACT TGT CTC CTG -3′.
3. Template TH2788 (*S. typhimurium* fliY5221::Tn*10*dTc) or any strain with a Tn*10*dTc insertion.

Note: When designing a complete deletion of a coding gene for allelic replacement with a *tetRA* element, leave the stop codon. For tetracycline-induced regulation of operons or genes, replace the promoter with the *tetRA* element.

Primer Design for tetRA *Elements with* rrnB T_1T_2 *Transcriptional Terminators and for* rrnB T_1T_2 tetRA

1. Primer rrnBT—25/10, 5′ 40-bps homology on sense strand + AGA GTA GGG AAC TGC CA-3′.
2. Primer tetA, 5′-40-bps homology on nonsense strand + CTA AGC ACT TGT CTC CTG-3′.
3. Template TH8094 *E. coli rrfB::tetRA.*

Note: rrnB T_1T_2 tetRA upon tetracycline induction will read out from *tetA.*

Primer Design for rrnB T_1T_2 tetAR2

1. Primer rrnBT—25/10, 5′-40-bps homology on sense strand + AGA GTA GGG AAC TGC CA-3′.
2. Primer tetA, 5′-40-bps homology on nonsense strand + CTA AGC ACT TGT CTC CTG -3′
3. Template TH8094 *E. coli rrfB::tetRA.*

Note: rrnB T_1T_2 tetRA upon tetracycline induction will read out from *tetA.*

Primer Design for rrnB T_1T_2 tetAR2

1. Primer rrnBT—25/10, 5'-40-bps homology on sense strand + AGA GTA GGG AAC TGC CA-3'.
2. Primer tet^{R2}, 5'-40-bps homology on nonsense strand + TTA AGA CCC ACT TTC ACA TT-3'.
3. Template TH8095 *E. coli rrfB::tetAR2*.

Protocol 1. Allelic Replacement or Insertion of the tetRA *Element into the Chromosome*

Day 1: Preparation of tetRA *Element Donor DNA*

1. PCR amplification of *tetRA* element flanked with homologous sequences for allelic replacement of insertion into the chromosome. Use primer sets and template DNA as described above.
2. PCR reactions of the tetRA elements flanked by 40 bps of homology are performed in a final volume of 50 μl in 1× PCR buffer, 200 μM dNTP, 0.5 μM of each primer, 100 to 200 ng genomic DNA, and one to five units of TAQ polymerase.
3. PCR reactions were performed at 95°, 3 min × 1 cycle; 95°, 30 sec, 49°, 30 sec, 72°, 2 min × 30 cycles, and 72°, 10 min × 1 cycle.
4. One-tenth the volume of the PCR reaction was checked on a 1% agarose-TAE gel (Sambrook and Russell David, 2001) to confirm amplifications of the 1990-bp flanked *tetRA* element. The PCR product was purified using QIAquick PCR purification Kit (Qiagen, Valencia, CA) according to the manufacturer's protocol except that the DNA was eluted in 25-μl EB buffer.
5. Determine DNA concentration.

Day 2: Preparation of Cells for λ-Red–Mediated Recombination

1. Grow TH4702 (pKD46 and *S. typhimurium* LT2) from a single colony in 50 ml LB + 100 μg/ml ampicillin at 30° until OD$_{600}$ ∼0.4 to 0.6. Add L-arabinose to a final concentration of 0.2% and induce for 1 hr.
2. Pellet cells at 10,000 × *g* for 5 min at 4°, and wash in 25 ml of cold sterile water. Pellet cells and repeat wash.
3. Re-suspend cells 1:10 volume (to 0.5 ml) in cold sterile water.
4. Electroporate 50 μl of cells, 100 to 200 ng of purified DNA of flanked *tetRA* in a 1-mm cuvette at 200 Ω, 1.6 kV, and 25 μF.
5. Immediately add 1 ml LB and incubate at 37° for 1 hr with shaking.
6. Plate 0.5 ml of cells on L-plates plus 12.5 μg/ml tetracycline and incubated at 37° overnight.

Note: The addition of L-arabinose induces the λ-Red genes *gam*, *bet*, and *exo* that are under the *araB* promoter on pKD46. In *E. coli*, constitutive expression of λ-Red has been reported to be mutagenic as shown by increased spontaneous resistance to rifampicin (Murphy and Campellone, 2003). However, it was found that there was no difference in spontaneous resistance to rifampicin in *E. coli* after 1 hr of induction of the λ-Red genes (Murphy and Campellone, 2003).

Day 3: Cure the pKD46 Plasmid from Putative Tc^R Colonies

1. Streak Tc^R colonies on L-plates without antibiotics and incubate at 42° overnight. This is to cure the strain of plasmid pKD46.
2. Confirm that colonies are Tc^R Ap^S.

Note: Using this protocol, the total numbers of Tc^R colonies isolated from 200 ng of flank *tetRA* element DNA is around 250 or 1.6×10^{-6} recombinants per survivor per microgram of DNA (Karlinsey, J. E., and Hughes, K. T., unpublished data).

Day Four: Confirm that the tetRA Element Was Integrated in the Correct Region on the Chromosome

1. Primer sets to ensure correct integration of the *tetRA* element include those from the *tetR* end. Design a sense strand primer of insertion downstream of the tetR and TetTOUT 5′-TAT TAC GAC AAG CTA TCG-3′ reads out 118 bp from stop codon of *tetR*. From the tetA end, T1Test 5′-TGC AGG AGA GAT TTC ACC GC-3′ reads out 750 bp from *tetA* stop, or Tn10-R1A 5′-AAT TGC TGC TTA TAA CAG GCA CTG-3′ reads out 158 bp from *tetA* stop and design a non-sense strand primer of insertion downstream of *tetA*.
2. Screen by colony PCR in the final volume of 20 μl in $1 \times$PCR buffer, 200 μM dNTP, 0.5 μM primer each, 0.2 units Promega TAQ polymerase/ Tc^R colony. The reactions were performed at 95°, 5 min, $\times$ 1 cycle; 95°, 30 sec; 45°, 30 sec; and 72°, 1 min, $\times$ 30 cycles.
3. Run PCR products on 1% agarose-TAE gel and check for predicted size products.

Note: If you plan to use the *tetRA* element for Tc^S counter-selection, check streak strains on Tc^S medium to make sure they do not grow.

Protocol 2. Allelic Replacement of tetRA Element Using Tc^S Counter-Selection for In Vivo Gene Construction of In-Frame Deletions, Gene Fusion, and Site-Directed Mutagenesis

Primer Designs for In-Frame Deletions, Gene Fusions, and Site-Directed Mutagenesis

1. Design and construct the tetRA element in the gene of interest to be deleted or at the specific DNA site to be targeted for site-directed mutagenesis.

2. Introduce λ-Red expression plasmid pKD46 by electroporation, selecting ampicillin resistance (100 μg/ml) on L medium at 30°.

3. Design a 80mer (up to 135mer can be made by IDT, Coralville, IA) oligonucleotide where the first half and the last half of the oligonucleotide are homologous to sequences upstream and downstream of the *tetRA* element, respectively. The 80mer can be designed to replace the *tetRA* element as an in-frame deletion of any size. For site-directed mutagenesis, the oligomer is designed with the site to be targeted in the middle of the 80mer. A 18mer complement to the 3′ end of the 80mer is also designed, which is hybridized to the 80mer followed by a fill-in reaction with DNA-polymerase Klenow fragment to make the oligomer double stranded.

4. For gene fusion constructions, design oligonucleotides to the fusion gene of interest (Gfp, His-tags, etc.), such that after PCR amplification they are flanked by 40 bps of homologies to the site of gene fusion.

5. Mutagenesis of genes or regions by *tetRA* replacement using donor DNA can be done with various methods. For example, with random or site-specific oligonucleotide, base replacement can be made (see step 3), and donor DNA can be made by error-prone PCR.

Day 1: Prepare Template DNA and Tetracycline-Sensitive (Tc^S) Medium

1. Prepare donor DNA with the considerations mentioned above.

2. Prepare tetracycline-sensitive medium as described (Maloy and Nunn, 1981). For 1-liter batch, prepare two flasks. Flask A contains 500 ml water, 15 g Bacto-Agar, 5 g Bacto-tryptone, 5 g yeast extract, and 50 mg chlorotetracyline. Flask B contains 500 ml water, 10 g NaCl, and 10 g NaH$_2$PO$_4$*H$_2$O (M.W. = 137.99). Autoclave 20 min, cool to 55°. To Flask B, add 5 ml Fusaric acid (2.4 mg/ml dissolved in dimethylformamide) and 5 ml 20 mM ZnCl$_2$. Mix Flasks A and B and pour into sterile petri plates. These plates are best when used fresh, but can be used if stored in foil at 4°.

Day 2: Preparation of Cells for λ-Red Mediated tetRA Replacement

1. Prepare the tetRA-containing strain with pKD46 for electroporation as described in Protocol 1, and plate dilution from 10^0 to 10^{-2} of cells onto Tc^S medium and incubate at 42° for 24 hr.

Note: It is important to include a cells-only control. This will allow you to compare the background of spontaneous Tc^S colonies that may arise with your parent strain. Typically, the number of Tc^S transformants is 100- to 1000-fold higher than the number of spontaneous Tc^S control colonies.

Days 3 and 4

1. Streak for isolated colonies once on Tc^S medium at 42°, and then on L-plates at 37°.

2. Check by PCR for the loss of the tetRA element, and confirm putative replacements by DNA sequencing.

References

Barbour, S. D., Nagaishi, H., Templin, A., and Clark, A. J. (1970). Biochemical and genetic studies of recombination proficiency in *Escherichia coli*. II. Rec+ revertants caused by indirect suppression of Rec-mutations. *Proc. Natl. Acad. Sci. USA* **67,** 128–135.

Cai, Y. P., and Wolk, C. P. (1990). Use of a conditionally lethal gene in *Anabaena* sp. strain PCC 7120 to select for double recombinants and to entrap insertion sequences. *J. Bacteriol.* **172,** 3138–3145.

Copeland, N. G., Jenkins, N. A., and Court, D. L. (2001). Recombineering: A powerful new tool for mouse functional genomics. *Nat. Rev. Genet.* **2,** 769–779.

Court, D. L., Sawitzke, J. A., and Thomason, L. C. (2002). Genetic engineering using homologous recombination. *Annu. Rev. Genet.* **36,** 361–388.

Daniels, D. W., and Bertrand, K. P. (1985). Promoter mutations affecting divergent transcription in the Tn*10* tetracycline resistance determinant. *J. Mol. Biol.* **184,** 599–610.

Datsenko, K. A., and Wanner, B. L. (2000). One-step inactivation of chromosomal genes in *Escherichia coli* K-12 using PCR products. *Proc. Natl. Acad. Sci. USA* **97,** 6640–6645.

Derbise, A., Lesic, B., Dacheux, D., Ghigo, J. M., and Carniel, E. (2003). A rapid and simple method for inactivating chromosomal genes in *Yersinia*. *FEMS Immunol. Med. Microbiol.* **38,** 113–116.

Hu, K., Shi, Z., Wang, H., Feng, E., and Huang, L. (2003). Study on gene knockout using red system in *Shigella flexneri*. *Wei Sheng Wu Xue Bao* **43,** 740–746.

Karlinsey, J. E., and Hughes, K. T. (2006). Genetic transplantation: *Salmonella enterica* serovar typhimurium as a host to study sigma factor and anti-sigma factor interactions in genetically intractable systems. *J. Bacteriol.* **188,** 103–114.

Karu, A. E., Sakaki, Y., Echols, H., and Linn, S. (1975). The gamma protein specified by bacteriophage gamma. Structure and inhibitory activity for the recBC enzyme of *Escherichia coli*. *J. Biol. Chem.* **250,** 7377–7387.

Kolodner, R., Hall, S. D., and Luisi-DeLuca, C. (1994). Homologous pairing proteins encoded by the *Escherichia coli recE* and *recT* genes. *Mol. Microbiol.* **11,** 23–30.

Lee, E. C., Yu, D., Martinez de Velasco, J., Tessarollo, L., Swing, D. A., Court, D. L., Jenkins, N. A., and Copeland, N. G. (2001). A highly efficient *Escherichia coli*-based chromosome engineering system adapted for recombinogenic targeting and subcloning of BAC DNA. *Genomics* **73,** 56–65.

Li, Z., Karakousis, G., Chiu, S. K., Reddy, G., and Radding, C. M. (1998). The beta protein of phage lambda promotes strand exchange. *J. Mol. Biol.* **276,** 733–744.

Little, J. W. (1967). An exonuclease induced by bacteriophage lambda. II. Nature of the enzymatic reaction. *J. Biol. Chem.* **242,** 679–686.

Liu, P., Jenkins, N. A., and Copeland, N. G. (2003). A highly efficient recombineering-based method for generating conditional knockout mutations. *Genome Res.* **13,** 476–484.

Maloy, S. R., and Nunn, W. D. (1981). Selection for loss of tetracycline resistance by *Escherichia coli*. *J. Bacteriol.* **145,** 1110–1111.

Murphy, K. C. (1991). Lambda Gam protein inhibits the helicase and chi-stimulated recombination activities of *Escherichia coli* RecBCD enzyme. *J. Bacteriol.* **173,** 5808–5821.

Murphy, K. C., and Campellone, K. G. (2003). Lambda Red-mediated recombinogenic engineering of enterohemorrhagic and enteropathogenic *E. coli*. *BMC Mol. Biol.* **4,** 11.

Murphy, K. C., Campellone, K. G., and Poteete, A. R. (2000). PCR-mediated gene replacement in *Escherichia coli. Gene* **246,** 321–330.

Muyrers, J. P., Zhang, Y., and Stewart, A. F. (2001). Techniques: Recombinogenic engineering–new options for cloning and manipulating DNA. *Trends Biochem. Sci.* **26,** 325–331.

Poteete, A. R. (2001). What makes the bacteriophage lambda Red system useful for genetic engineering: Molecular mechanism and biological function. *FEMS Microbiol. Lett.* **201,** 9–14.

Poteete, A. R., Fenton, A. C., and Nadkarni, A. (2004). Chromosomal duplications and cointegrates generated by the bacteriophage lamdba Red system in *Escherichia coli* K-12. *BMC Mol. Biol.* **5,** 22.

Sambrook, J., and Russell David, W. (2001). "Molecular Cloning: A Laboratory Manual." Cold Spring Harbor Laboratory Press, Cold Spring Harbor, NY.

Stahl, M. M., Thomason, L., Poteete, A. R., Tarkowski, T., Kuzminov, A., and Stahl, F. W. (1997). Annealing vs. invasion in phage lambda recombination. *Genetics* **147,** 961–977.

Uzzau, S., Figueroa-Bossi, N., Rubino, S., and Bossi, L. (2001). Epitope tagging of chromosomal genes in *Salmonella. Proc. Natl. Acad. Sci. USA* **98,** 15264–15269.

Way, J. C., Davis, M. A., Morisato, D., Roberts, D. E., and Kleckner, N. (1984). New Tn*10* derivatives for transposon mutagenesis and for construction of *lacZ* operon fusions by transposition. *Gene* **32,** 369–379.

Yamaguchi, A., Udagawa, T., and Sawai, T. (1990). Transport of divalent cations with tetracycline as mediated by the transposon Tn*10*-encoded tetracycline resistance protein. *J. Biol. Chem.* **265,** 4809–4813.

Yu, D., Ellis, H. M., Lee, E. C., Jenkins, N. A., Copeland, N. G., and Court, D. L. (2000). An efficient recombination system for chromosome engineering in *Escherichia coli. Proc. Natl. Acad. Sci. USA* **97,** 5978–5983.

Zhang, Y., Buchholz, F., Muyrers, J. P., and Stewart, A. F. (1998). A new logic for DNA engineering using recombination in *Escherichia coli. Nat. Genet.* **20,** 123–128.

[17] Probing Nucleoid Structure in Bacteria Using Phage Lambda Integrase-Mediated Chromosome Rearrangements

By Nathalie Garcia-Russell, Samantha S. Orchard, and Anca M. Segall

Abstract

Conservative site-specific recombination has been adapted for a multitude of uses, in both prokaryotes and eukaryotes, including genetic engineering, expression technologies, and as probes of chromosome structure and organization. In this article, we give a specific example of the latter application, and a quick summary of some of the myriad other genetic and biotechnology applications of site-specific recombination.

METHODS IN ENZYMOLOGY, VOL. 421   0076-6879/07 $35.00
                                                    DOI: 10.1016/S0076-6879(06)21017-6

Introduction

The structure of the *E. coli* and *Salmonella* chromosomes has been under investigation for over 30 years, using a variety of methods including electron microscopy (Kavenoff and Bowen, 1976; Robinow and Kellenberger, 1994), sedimentation centrifugation (Dworsky and Schaechter, 1973; Hecht *et al.*, 1977; Sinden and Pettijohn, 1981), expression pattern analysis (Schmid and Roth, 1987), transposition and site-specific recombination (Garcia-Russell, *et al.*, 2004; Krug *et al.*, 1994; Manna and Higgins, 1999; Valens *et al.*, 2004), and more recently, bioinformatics (Capiaux *et al.*, 2001; Hendrickson and Lawrence, 2006; Levy *et al.*, 2005; Ussery *et al.*, 2001), pulsed field gel electrophoresis (Liu and Sanderson, 1996), site-specific recombination (Deng *et al.*, 2005), and fluorescence *in situ* hybridization (reviewed in Gordon and Wright, 2000; Nielsen *et al.*, 2006; Niki *et al.*, 2000; Viollier *et al.*, 2004; Wang *et al.*, 2006). Different methods have probed and revealed various levels of organization. Several limitations have made a complete understanding of the chromosome elusive, chief among them the small size of bacteria and the limit of resolution of light microscopy. The advancement of deconvolution microscopy coupled with the use of fluorescently labeled DNA-binding proteins has permitted much more incisive analysis of the position of various chromosomal regions with respect to the growth phase (e.g., Bates and Kleckner, 2005). Since the 1980s, two books (Higgins, 2005; Charlebois, 1999) and a large number of reviews have covered a wide range of topics related to bacterial chromosomes, but the issue has not yet been solved. This article outlines our use of one of these methods, site-specific DNA rearrangements catalyzed by the bacteriophage lambda integrase protein, to probe the ability of different regions of the bacterial chromosome to come into close-enough contact to be successfully paired and enzymatically transformed. Given data on a sufficient number of points, this method should give us a three-dimensional view of the folded structure of the chromosome in wild-type cells and mutants in a number of the nucleoid-associated proteins characterized in bacteria to date.

What Is Known About Chromosome Structure

Just as in eukaryotic chromosomes, several levels of organization have emerged in bacterial chromosomes, including what appear to be six macrodomains, defined by the ability to perform recombination between sites within the macrodomain but not outside it (Valens *et al.*, 2004), and stochastic short 10- to 12-kb domains established by transcription activity (Deng *et al.*, 2005; Postow *et al.*, 2004). Despite the fact that bacterial DNA is not tightly wrapped into chromatin, about 50% of it is constrained by

DNA-binding proteins (Bliska and Cozzarelli, 1987). Some of these, such as HU, IHF, H-NS, and Fis, are very abundant—5000 up to 30,000 proto-mers per cell, depending on the growth phase (Azam *et al.*, 1999; Johnson *et al.*, 2005).

Several DNA-transacting proteins have been adapted as probes of chromosome organization. The ability of transposases to access various areas of the chromosome has revealed unequal distribution of transposon insertion sites (Krug *et al.*, 1994), although the sequence preference of transposases resulting in hot spots and cold spots (Manna *et al.*, 2004) and transposition immunity phenomena (Darzins *et al.*, 1988; DeBoy *et al.*, 1996) have made transposon data harder to interpret as a simple reflection of chromosome structure. Site-specific recombination experiments have pro-duced more easily interpretable results. These experiments have been done with two different site-specific recombinases, the $\gamma\sigma$ resolvase (Res), and phage lambda integrase (Int). The mechanistic differences between the two systems (for an overview of mechanistic details, see Azaro and Landy, 2002, and Grindley, 2002) are such that the two enzymes operate on different "geographic" scales, with the Res protein being restricted to recombination over relatively short ranges (in general spanning no more than about 100 kb or 2% of the *Escherichia coli* or *Salmonella* chromosome), while Int is able to catalyze recombination between sites separated by half of the bacterial chromosome (Garcia-Russell *et al.*, 2004). In part, these restrictions are due to the mechanistic distinctions between the two enzymes—Res must pair its recombination substrates in a specific topological arrangement with respect to each other (Watson *et al.*, 1996), and is thus restricted to recombining sites within the same "topological domain." Int on the other hand does not have topological restrictions on pairing of its recombination substrates and can recombine sequences that are very far apart either on linear or super-coiled DNA molecules, or even on two different DNA molecules, as long as the sites can interact in three-dimensional space (Nash and Pollock, 1983). (In fact, Int evolved to recombine sequences on two different DNA mole-cules, the phage genome and the bacterial genome, while Res evolved to recombine sequences present on the same genome to generate a deletion between two recombination sites.) However, restrictions are also imposed by the recombination screen itself; a screen that demands growth and thus viability will not be able to uncover those bacteria that have a lethal rear-rangement (e.g., a deletion of a chromosomal segment that contains essen-tial genes or an inversion that is nonpermissive for bacterial growth) (Rebollo *et al.*, 1988; Segall *et al.*, 1988).

In this chapter we describe a PCR-based screen that does not demand bacterial growth, and simply reports, by inference, the ability of the Int protein to physically pair sites as a function of their position in the chromosome.

An advantage of this screen is that the bacteria that have undergone recombination do not need to be alive. This is a significant worry in the case of the intrachromosomal recombination reactions (see below), which may generate a very substantial and possibly lethal rearrangement. In the intermolecular assay between a plasmid and the chromosome, recombination results in the plasmid being integrated into the chromosome. High-copy-number plasmids are very unstable when integrated in this situation; for that reason, we have placed the *att* sites on a low-copy-number plasmid that should be well tolerated.

We have designed a probe of chromosome structure based on the site-specific recombination system encoded by phage lambda (Garcia-Russell *et al.*, 2004) (Fig. 1). The phage encodes a recombinase, which synapses, cuts, and re-ligates pairs of specific sequences named *att* sites. Recombination between the *att* sites is limited by the ability of the two sites to be synapsed, and thus the extent of recombination reports the ability of the two regions where the *att* sites are placed to come into close proximity. We have used this assay to test recombination either between two chromosomal loci (intramolecular) or between one chromosomal locus and a plasmid (intermolecular) (Garcia-Russell *et al.*, 2004). In general, intermolecular recombination occurs at higher frequency than intramolecular recombination, possibly because the plasmid (and therefore the *att* site) has a higher copy number than the chromosome (about six copies per cell), and possibly because the plasmid may be subject to fewer constraints on its movement. The Int and Xis proteins can be supplied from genes expressed on compatible plasmids, on a single plasmid, or integrated in the chromosome. The expression of the *int* gene in particular should be tightly repressed to avoid toxicity to the host cell but rapidly inducible to allow fast recombination and its assessment to occur before the potentially harmful effects of the recombination affect the detection of the product (i.e., if the rearrangement is lethal, the cells harboring it will die and be outcompeted in the population). For this purpose, we found that the *lacUV5* promoter was not sufficiently tightly regulated, particularly on plasmids such as pBR322 derivatives. (In such a situation, the uninduced *int* gene was expressed sufficiently to support recombination after overnight growth of the culture [T. Q. Le, Master's thesis, San Diego State University, 1999].) On the other hand, expression of both *int* and *xis* from the *pBAD* promoter requires at least a 2-h induction time before recombination is detected in a gel-based assay (Procedure 1 below [K. Karrento, N. Garcia-Russell and A. Segall, unpublished results]). We have found that a *lac* promoter with two LacI operators (Lanzer and Bujard, 1988) is sufficiently repressed to see IPTG-dependent recombination in almost all situations we have assayed, but permits easily measurable recombination in 30 min (15 min is generally

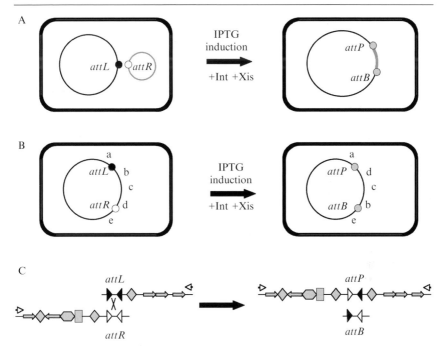

Fɪɢ. 1. Schematic of the site-specific recombination assay for probing chromosome structure. (A) Intermolecular recombination. One *att* site is on the chromosome while the partner *att* site is on a low copy–number plasmid. When Int and Xis expression is induced, recombination occurs and the plasmid is integrated into the bacterial chromosome. (B) Intramolecular recombination. Both *att* sites are in the chromosome, and recombination between them generates an inversion of the intervening segment of DNA. (C) Details of the four different *att* sites. Strand exchange occurs in the region between the horizontal black and white triangles, which represent the Int core-binding sites. Recombination re-sorts sequences to create hybrid *att* sites, as shown. The gray shapes indicate binding sites for the recombination accessory proteins (host-encoded IHF and Fis, and phage-encoded Xis) (Azaro and Landy, 2002), as well as for the accessory domain of Int protein known as the arm-binding domain. The amplification primers are denoted by the open arrowheads.

not sufficient except for the most "permissive" pair of recombination substrates). Recombination between an *attL* and an *attR* site will generate two unique recombination products, and either can be assayed. We have generally assayed the *attP* product because it is larger than the *attB* product, and use the *attL* substrate as a control.

(Note that we use *attL* X *attR* recombination rather than *attP* X *attB* recombination. There are several reasons for this practice, including the two mentioned here. First, this recombination is less sensitive to the local superhelical density environment, since recombination between *attL* and *attR* can occur quite well when both sites are either supercoiled or relaxed, or when one is supercoiled and the other relaxed; and second, both *E. coli* and *Salmonella* have an *attB* site appropriate for phage lambda, which would have to be deleted so as not to interfere with *attP* X *attB* recombination elsewhere. In contrast, neither *attL* nor *attR* recombine appreciably with *attB*, and thus the endogenous site does not have to be deleted, and may be used as a reference.)

PCR products can be detected and measured using either a semiquantitative gel-based method (Procedure 1) or a quantitative real-time PCR (qPCR) method (Procedure 2). The gel-based method is less costly and does not require any special equipment except a "garden-variety" PCR machine, an electrophoresis set-up, and a digital camera for capturing the gel image. We use acrylamide gel electrophoresis because the greater transparency of polyacrylamide gels makes them more sensitive to quantitate than agarose gels. The gel-based method has the drawback that it is low through-put (running four gels every other day with 20 lanes each is the maximum reasonable), and has a more limited sensitivity (roughly one recombinant chromosome in 2000 cells). In principle, qPCR is significantly more sensitive, and we have been able to detect 1 cell in 100,000 that has been rearranged. Moreover, the manipulations are less onerous and one or two 96-well plates could be analyzed per day, if necessary. Finally, qPCR is by its nature more easily quantitated.

Experimental Outline

First, the appropriate strains to test recombination must be constructed. The insertion of *att* sites can be done by the recombineering/Red Swap method or by constructing transposons carrying the *att* sites and allowing their transposition around the chromosome. While the Xis gene can be present during the construction of the strain with the two *att* sites, the Int gene should be added last, if possible, before recombination is assayed.

Second, the strains for testing recombination must be grown as appropriate for the experiment; in the present case, strains are subcultured to achieve exponential growth as measured by turbidity ($OD_{600} = 0.4$). The recombination proteins are then induced for 30 min. After inducing recombination, two possibilities exist: (1) the cultures are used immediately as templates for the PCR assay, or (2) genomic DNA can be isolated and then used as a template. In the first case, recombination can be expressed as a number of recombined genomes per million cells. In the latter, recombination can be expressed as recombined genomes per nanogram (ng) of DNA. A variety of controls must be performed as appropriate for each procedure.

Procedures

Procedure 1. Gel-Based Assay for Measuring Chromosome Fluidity

Strains

Strain with one *attL* site and one *attR* site (each is marked with a drug-resistance gene for ease of strain construction), with the *int* and *xis* genes downstream of the Bujard promoter; for each experiment, 6 to 12 independent colonies are used.

Strain with *att* sites but without the *int* gene.

Strain with a single-copy *attP* site in the chromosome to be used as a control (e.g., SDT1287, *sicA::attP*).

Primers

Primer 1 (attRXho) 5'- CCG CTC GAG GCG CTA ATG CTC TGT TAC AG

Primer 2 (P'29BamHI) 5'- CGC GGA TCC CAG GGA GTG GGA CAA AAT TG

Primer 3 (Neo Middle) 5'- TAA TGG CTG GCC TGT TGA AC

Other Supplies

Clear, flat-bottomed 96-well microtiter plates for growing cultures; the OD_{600} can be read directly in a plate reader.

Media with appropriate antibiotics. Once strains are constructed, plasmids should be selected continuously and antibiotic resistances associated with the recombination proteins should also be selected, but the resistances associated with chromosomal *att* sites do not need to be selected continuously.

dNTP mix (e.g., Bioline); Klentaq1 Polymerase (AB Peptides or Klentaq.com); $10\times$ PC2 (Klentaq1) buffer and 10-100X Mg^{2+} stock solution and appropriate primers (Garcia-Russell *et al.*, 2004).

Preparation

1. Inoculate 6 to 12 independent colonies of the tester strains and a control strain without the Int gene into a microtiter plate with 100 μl of media per well. To reduce the possibility of evaporation, do not inoculate the wells around the edge of the plate unless you are confident that you can seal the plate completely with sealing film. Grow overnight, add sterile glycerol to a final concentration of 12.5%, and freeze the plate at $-80°$. This plate can be used as the source of cells for several experiments.

2. Before an experiment, inoculate a fresh microtiter plate by using a multichannel pipettor with tips to scrape up some of the cell ice as the inoculum. Keep the source plate on ice while inoculating cultures and

return to the −80° freezer as soon as possible. Incubate the inoculated plate overnight at an appropriate temperature in an air-shaker by taping the plate on the platform or on a test-tube rack. In all multichannel manipulations, make sure that all the tip volumes are the same. Organize similar treatments in either rows or columns, for ease of handling.

3. The next day, dilute the overnight cultures 1:10 or 1:20 into fresh media with appropriate antibiotics. Use a multichannel pipettor, sterile reservoirs (sterilizable ones are available from Fisher Scientific), and sterile aerosol-resistant tips. Pipet up and down to mix solutions in the source plate and in the target plate after inoculation. It is extremely important to avoid all possible contamination. Check OD_{600} to ensure equal inocula.

4. IPTG induction. After appropriate length of time for the genetic background being tested, check OD_{600} (∼4 h). When it reaches the appropriate OD, withdraw up to 5 μl for viable cell counts into a microtiter plate in which diluent such as 10 mM Tris/1 mM Mg^{2+} or phosphate-buffered saline (PBS) has been added, and keep this on ice. Add IPTG solution (dissolved in Luria broth [LB] or a minimal media) to achieve a final concentration of 0.5 mM. Incubate for 30 min at 37°, 250 rpm.

5. During the initial incubation to bring cultures into exponential phase, prepare the PCR reaction mix (50 μl per well). *Precautions*: Use only aerosol-resistant tips to avoid contamination; dilute dNTPs and primers in sterile water kept only for this purpose. Change this water source frequently. Keep pre-mixes on ice, or make the pre-mix without enzyme the previous day and freeze at −20°. Always add the enzyme fresh (see Table I).

TABLE I
PCR REACTION MIX FOR THE GEL-BASED ASSAY

Pre-mix components	Volume per reaction[a]
Primer 1 (20 μM stock)	2 μl
Primer 2 (20 μM stock)	4 μl[b]
Primer 3 (20 μM stock)	2 μl
10× PC2 buffer[c]	5 μl
dNTPs (20 mM stock)	0.5 μl
Klentaq1	0.4 μl
Sterile nH$_2$O	31.1 μl

[a] Prepare an extra reaction for every 15 or more reactions to allow for pipetting errors. Beware of surface tension when pipetting.

[b] In the gel-based method, we use three primers: one is specific for the *attP* product, one is specific for the *attL* substrate, and the third is common to *attP* and *attL*. We add twice as much of this third primer, since it is often limiting otherwise.

[c] We have found that the Klentaq1 polymerase (Barnes, 1994) gives us the most robust and reproducible signal among the polymerases we have tested when using cell culture directly as the template. The 10× PC2 buffer comes with the enzyme.

When assembling the premix, always use a fresh sterile reservoir. Also, for each gel, assemble a reaction that has an *attP* plasmid (very dilute, 10 to 20 ng) to use as the positive control.

6. Distribute 45 μl PCR mix into each tube or well, and then add 5 μl of the appropriate culture and mix well by pipetting up and down. Pay attention to the uniformity of the volumes in the multichannel pipette tips. Spin the plate or tubes briefly ($\sim$10 s, at no more than 500 rpm to avoid pelleting the bacteria) in a refrigerated centrifuge to bring the liquid to the bottom of the wells and to eliminate air bubbles.

Place plate in a 96-well thermocycler when the temperature has reached at least 80°, and use the hot-top feature (or add a drop of mineral oil to each reaction). We use a touch-down PCR program, consisting of an initial 5-min denaturation step at 95°, 20 to 25 cycles of 1 min denaturation at 95°, 1 min annealing at 72° and decreasing by 2° every two cycles, and terminating with 15 cycles at 62°, and 1 min of elongation at 72°, with a final elongation step of 5 min. However, the temperature profile and the cycle number should be optimized for each set of primers. Keep the plate at 4° after the end of the cycling program prior to loading the gel.

7. PAGE analysis of the PCR products: The PCR reactions are analyzed on a Tris-Tricine-SDS gel with a final 5% polyacrylamide concentration (Table II). We have found this gel system optimal for giving us the least smeared and most easily quantitated bands.

Pour gel using 1.5-mm spacers with 20-well combs. Use 1× Tris-tricine-SDS running buffer (10× stock solution: 121 g Tris base, 179 g tricine, 10 g SDS). Load entire 50-μl reaction after mixing with 10 μl of a 6× Ficoll-based loading dye (Sambrook *et al.*, 1989). Run gel for about 5 h at a constant 100 mA current. The dye migrates with the salt front—stop gel when dye is at the bottom or just off of the gel.

8. Ethidium bromide staining. Rinse the SDS out of the gel in 200 ml of distilled or reverse osmosis-treated H_2O by gently shaking on a rotating platform for 30 min. Stain gel with 5 μl of 10 mg/ml EtBr stock in 100 ml of water. De-stain gel in 200 ml of water. See Fig. 2 for a sample of the data obtained in a gel-based assay.

9. Develop a standard procedure for taking a picture of the gel. This is extremely important in order to be able to compare gels from multiple days of experiments. We use a Stratagene EagleEye image capture system, with the camera aperture set at 2.8.

10. Analysis of the data. The gel image is saved on a disk and analyzed using the ImageQuant program that comes with Storm or Typhoon phosphorimagers (GE Molecular Dynamics); alternatively, the free ImageJ software made available by the National Institutes of Health (http://rsb.info.nih.gov/ij/) can be used to measure pixel density of the *attP* and the *attL*

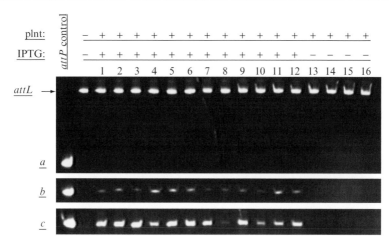

Fig. 2. Examples of the gel-based method for detecting recombinant products. The results shown are for three independent chromosomal intervals showing undetectable recombination in the case of *a*, low recombination in the case of *b*, and high recombination in the case of *c* (the *attP* products are next to the letters *a*, *b*, and *c*). Gels *b* and *c* were cropped to show only the *attP* product. (Adapted from Garcia-Russell *et al.*, 2004.)

products. In principle, the amounts of *attP* and *attL* should add to 100% in each gel lane. In practice, this rarely works well because there is often overwhelmingly more *attL* substrate than *attP* product. Instead, we normalize the *attP* values to the OD_{600} values (more convenient than colony-forming units per milliliter for all the replicates). Statistical analysis is then performed, including analysis of covariance (ANOVA) to determine whether a global effect is evident, followed by the Tukey's test to compare the means. Other tests may be appropriate depending on the comparisons necessary (e.g., we have used the Student's *t*-test to analyze differences between mutant bacteria).

Procedure 2. qPCR Measurement of Chromosome Fluidity

The qPCR assay we have developed uses Molecular Beacons rather than TaqMan probes to monitor fluorescence (Fig. 3). (For a recent review of available real-time PCR quantitation methods, see Wong and Medrano, 2005.) The TaqMan assay depends on the 5′- to 3′-nuclease activity of the Taq polymerase. Because we have had more consistent results with Klentaq1 when using bacterial cultures and this enzyme lacks this nuclease activity (Barnes, 1994), we cannot use the TaqMan system. In principle, using

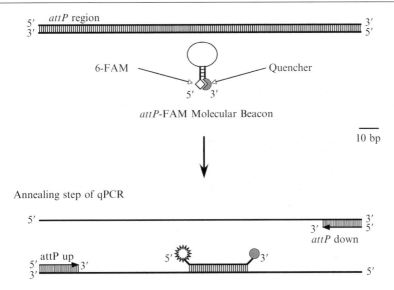

Fɪɢ. 3. qPCR detection of *attP* recombinant products. Chromosomes that have recombined will have an *attP* site, which will be subjected to PCR during the qPCR assay, as shown in Fig. 1 (except that a product of only ~100 to 150 bp is desired during the qPCR assay). The Molecular Beacon is closed until the denaturation stage of the PCR reaction. During denaturation, the Molecular Beacon will open. If a complementary *attP* substrate is present in the tube, the Molecular Beacon can anneal to it. In fact, the Beacon is designed so that annealing to its appropriate substrate is favored over re-closing the stem loop. The Molecular Beacon fluoresces when it is in the open conformation (i.e., annealed to substrate), and is quenched in the closed conformation. The more *attP* substrate is available, the higher the number of open Molecular Beacons and the more intense the fluorescence. During elongation, the Beacon will be dissociated by the elongating polymerase, but fluorescence is measured during the annealing step of every PCR cycle. The Molecular Beacon was designed to anneal to the core region of the *attP* site, taking advantages of differences between *attL* and *attP*.

Molecular Beacons also requires less double-labeled oligomer probe in each assay, since the TaqMan probes are consumed in the assay, while Molecular Beacons are reusable in every cycle. On the other hand, Molecular Beacons may be somewhat harder to optimize initially, particularly in a situation where two or more would be used simultaneously for multiplex applications (N. Garcia-Russell, unpublished results). In practice, both assays could be adapted for this purpose in cases where purified DNA is used as the template.

Strains

LT2 wild-type *Salmonella enterica* serovar Typhimurium
SDT1287 LT2 *sicA::attP*
Tester strains containing pairs of *att* sites, Int and Xis

Primers

> attPup: TTA CAG TAT TAT GTA GTC TG
> attPdown: TTT TGA CTG ATA GTG ACC TG
> Molecular Beacon 5′/6-FAM/CGC GTC GTT CAG CTT TTT TAT
> ACT AAG TTG-GCA CGC G/BHQ_1/3′ (37 nt)

Preparation

1. This protocol can either start with steps 1 to 6 in Procedure 1, followed directly by the qPCR analysis of the PCR products, or by growing liquid cultures in tubes rather than microtiter plates and isolating chromosomal DNA. If using purified DNA templates, start 2-ml cultures from 4–6 independent colonies of the appropriate tester strains. Use the appropriate media supplemented with antibiotics (maintain selection for the *int* and *xis* gene insertions, whether chromosomal or on a plasmid, since they can be unstable). In addition, start a culture of LT2 and of SDT1287 (a strain containing *attP* on the chromosome). LT2 will serve as the negative control. SDT1287 will be used to make a standard curve for the qPCR.

2. Subculture each of the cultures (including LT2 and SDT1287) 1:20 into 3.5 ml of LB. Place 150 μl into a microtiter plate and take the absorbance of the culture at OD_{600}. Incubate the cultures at 37° in a shaking incubator. If using cells rather than purified DNA as PCR templates, the cultures can be subcultured into a total volume of 150 μl in a microtiter plate and the same plate may be used both for optical density readings and PCR inocula.

3. Follow the absorbance every hour until the culture gets near OD_{600} ∼ 0.3, and then monitor every 30 min or more often, depending on how close they are to 0.4. If using purified DNA, harvest 1 ml of each culture at OD_{600} = 0.4 and keep on ice until all are ready for DNA isolation. From each of the 1-ml cultures on ice, isolate chromosomal DNA using an appropriate protocol or kit (e.g., the Qiagen DNeasy Tissue Isolation Kit). If adding cultures directly into PCR reactions, add 2.5 μl of culture immediately to a pre-prepared microtiter plate.

4. Quantitate purified DNA, if applicable, by taking the OD_{260}/OD_{280} readings.

5. Assemble PCR reactions using the recipe. Assemble an appropriate pre-mix as necessary, and then apportion into 0.2-ml tube strips or microtiter plates, depending on the number of reactions being done. Keep the pre-mix on ice until ready to start the PCR, and then add the template DNA. Use 100 ng of DNA per reaction. Reactions using the genomic DNA preparations of LT2 and SDT1287 should be assembled as negative and positive controls.

TABLE II
RECIPE FOR TRIS-TRICINE-SDS POLYACRYLAMIDE GELS

Gel recipe (for ~15 cm × 15 cm gel)	Volume
3M Tris-Cl, pH 8.45	16.7 ml
nH₂O	20 ml
30% polyacrylamide stock (29:1)[a]	8.3 ml
50% glycerol (50g/100 ml)	0.5 ml
10% SDS	0.5 ml
10% ammonium persulfate	0.34 ml
TEMED (**add last**)	75 μl

[a] 29 g acrylamide: 1 g bis-acrylamide per 100 ml.

6. Start the qPCR run as prescribed for the instrument available. We use an MJ Research Chromo4 instrument (now sold by BioRad), which we find to be the most sensitive, a particularly important consideration when cultures rather than genomic DNA preparations are used as templates.

7. Analysis of assay results. When the real-time PCR run is completed, the data must be verified and then the calculated amount of template must be converted into meaningful and easily understood units. If the computer has automatically set a threshold level (the level of fluorescence where the curves are above background but not out of exponential range), verify its position (Fig. 4). If the computer has not set the threshold, it must be set manually. For this the fluorescence axis must be set in log scale in order to identify the period of the PCR where product was increasing exponentially relative to cycle number. In log scale, the threshold should pass through the data lines where as many of them as possible are straight. After choosing an appropriate position for the threshold, the standard curve must be verified (determined using the data for the serial dilutions of the *attP*-carrying strain). The software most likely provides a straight-line equation for the standard curve (log of standard template quantity versus cycle threshold number) as well as an r^2 value to indicate the degree to which the data points of your standards align with the straight-line equation. The slope of the line (m, in $y = mx + b$) should be close to -0.302 on a graph of the log of standard quantity plotted against cycle threshold number, or -3.32 for a graph where the axes have been reversed; a slope near these values reflects a perfect doubling of product with each cycle. The r^2 value should be higher than 0.9 but the closer it is to 1, the better. Once the threshold has been set and the standard curve has been adjusted, ask the software to calculate the amount of starting template for all the samples. It will do so by computing the cycle number at which the sample data curves cross the threshold through the straight-line equation of the standard curve (the Ct value).

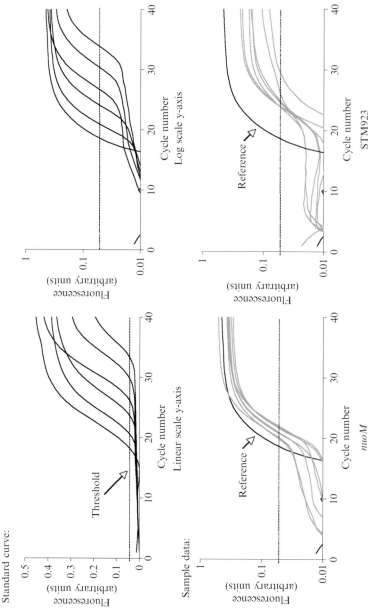

Fig. 4. Example of qPCR data. The top two panels show a standard curve displayed on a linear scale (left) and on a log scale (right). The higher the concentration of DNA substrate being detected, the fewer cycles it takes for the curve to cross the threshold. Although in this case the threshold line can be set on both graphs more or less equally well, it is important to check the threshold when the y-axis is on a log scale. In the bottom two panels, recombination assay data from two different strains are shown. In each case, the reference curve shown is from the highest concentration of DNA of an *attP*-containing strain, while the other lines are independent

TABLE III
PCR REACTION MIX FOR THE qPCR ASSAY

Reaction components	1 Reaction	25-Reaction pre-mix
Primer 1 (15 μM)	0.5 μl	12.5 μl
Primer 2 (45 μM)	0.5 μl	12.5 μl
dNTPs (10 mM)	0.5 μl	12.5 μl
PC2 buffer (10×)	2.5 μl	62.5 μl
attP-FAM (25 μM)	0.5 μl	12.5 μl
Klentaq1	0.25 μl	6.25 μl
H$_2$O	17.75 μl	444 μl
template DNA (100 ng)	2.5 μl	
Final volume	25 μl	

The units of your data will depend on the units of your standards: when using whole cells for both standards and samples, the units are the "copy number" of the expected product, while for purified DNA the units are nanograms of *attP* product DNA. If using whole cells, the colony-forming units of serial dilutions of the positive control (*attP*) culture are calculated and compared to the amount (or number of copies) of *attP* obtained. Using this standard curve, the results of the qPCR for the experimental strains can be converted to meaningful units, such as "recombination events per 10^6 CFU" or "percent of cells in which recombination occurred" (Table III).

Comment: Genetic Engineering Using Phage Integrases

Many bacteriophages establish and maintain stable lysogeny relationships with their hosts, frequently integrating into their hosts' genomes, and their genome is replicated passively along with their hosts'. The integrases have evolved to recombine the phage genome into the bacterial one, at specific sequences, in a conservative manner such that no sequences are gained or lost. Thus, the integrity and genetic information of both genomes are maintained after the rearrangement. This fidelity of the exchange reaction makes these enzymes very useful for genetic engineering. For example, the chromosomal *attB* site for phage lambda is very convenient for the integration of genetic material in a single copy such as a gene, wild-type or mutant, for complementation purposes or reporter fusions (e.g., Diederich *et al.*, 1992). More recently, the methods of recombineering and Red Swap for making targeted

experiments of the tester strain shown (usually at least four independent cultures or DNA assayed in duplicate). The data shown were obtained by students in the Cold Spring Harbor 2006 Advanced Bacterial Genetics Course. (Data for recombination at the *nuoM* locus were obtained by Tao Long and Caroline Roper, while data for recombination at the STM923 were obtained by Jennifer Carr and Lindsay Wilson.)

deletions or mutations in the chromosomes of some enteric bacteria have greatly facilitated the engineering of complex strains (see Chapter 15), but more and more we are running out of drug-resistance genes to use as effective markers for genetic selections. In such cases, the conservative site-specific deletion of antibiotic-resistance genes using a relative of Int, the yeast Flp protein encoded on the 2 μ circle (Jayaram *et al.*, 2002), has been extremely useful (Baba *et al.*, 2006; Datsenko and Wanner, 2000; Palmeros *et al.*, 2000). Another terrific use for phage integrase-mediated reactions is to achieve extremely tight control of gene expression, by providing a promoter for a specific (toxic) gene only after inverting a promoter pointed the wrong way to its correct orientation via site-specific recombination (Sektas *et al.*, 2001; Sektas and Szybalski, 1998). Finally, the lambda Int-based Gateway system (InVitrogen Corp.) (Walhout *et al.*, 2000) allows cloning-free shuffling of the same open reading frame among a number of vectors to permit expression in various systems and also allows fusions of proteins to different tags for purification purposes.

Conclusion

Here we have skimmed the surface of the versatility of site-specific recombination as a tool for a variety of purposes. More of these conservative recombination systems have been and will continue to be uncovered in their native organisms and adapted for use, and their utility will continue to expand. The needs and creativity of molecular biologists will fuel more, and improved, applications.

References

Azam, T. A., Iwata, A., Nishimura, A., Ueda, S., and Ishihama, A. (1999). Growth phase dependent variation in protein composition of the *Escherichia coli* nucleoid. *J. Bacteriol.* **181,** 6361–6370.

Azaro, M. A., and Landy, A. (2002). λ Integrase and the λ Int family. *In* "Mobile DNA II" (N. L. Craig R. Craigie M. Gellert and A. M. Lambowitz, eds.). ASM Press, Washington, DC.

Baba, T., Ara, T., Hasegawa, M., Takai, Y., Okumura, Y., Baba, M., Datsenko, K. A., Tomita, M., Wanner, B. L., and Mori, H. (2006). Construction of the *Escherichia coli* K-12 in-frame, single gene knockout mutants: The Keio collection. *Mol. Syst. Biol.* **2,** 2006–2008.

Barnes, W. M. (1994). PCR amplification of up to 35-kb DNA with high fidelity and high yield from lambda bacteriophage templates. *Proc. Natl. Acad. Sci. USA* **91,** 2216–2220.

Bates, D., and Kleckner, N. (2005). Chromosome and replisome dynamics in *E. coli*: Loss of sister cohesion triggers global chromosome movement and mediates chromosome segregation. *Cell* **121,** 899–911.

Bliska, J. B., and Cozzarelli, N. R. (1987). Use of site-specific recombination as a probe of DNA structure and metabolism *in vivo*. *J. Mol. Biol.* **194,** 205–218.

Capiaux, H., Cornet, F., Corre, J., Guijo, M. I., Perals, K., Rebollo, J. E., and Louarn, J. M. (2001). Polarization of the *Escherichia coli* chromosome. A view from the terminus. *Biochimie* **83,** 161–170.

Charlebois, R. L. (ed.) (1999). "Organization of the Prokaryotic Genome." ASM Press, Washington, DC.

Darzins, A., Kent, N. E., Buckwalter, M. S., and Casadaban, M. J. (1988). Bacteriophage Mu sites required for transposition immunity. *Proc. Natl. Acad. Sci. USA* **85,** 6826–6830.

Datsenko, K. A., and Wanner, B. L. (2000). One-step inactivation of chromosomal genes in *Escherichia coli* K-12 using PCR products. *Proc. Natl. Acad. Sci. USA* **97,** 6640–6645.

DeBoy, R. T., and Craig, N. L. (1996). Tn7 transposition as a probe of cis interactions between widely separated (190 kilobases apart) DNA sites in the *Escherichia coli* chromosome. *J. Bacteriol.* **178,** 6184–6191.

Deng, S., Stein, R. A., and Higgins, N. P. (2005). Organization of supercoil domains and their reorganization by transcription. *Mol. Microbiol.* **57,** 1511–1521.

Diederich, L., Rasmussen, L. J., and Messer, W. (1992). New cloning vectors for integration in the lambda attachment site *attB* of the *Escherichia coli* chromosome. *Plasmid* **28,** 14–24.

Dworsky, P., and Schaechter, M. (1973). Effect of rifampin on the structure and membrane attachment of the nucleoid of *Escherichia coli*. *J. Bacteriol.* **116,** 1364–1374.

Garcia-Russell, N., Harmon, T. G., Le, T. Q., Amaladas, N. H., Mathewson, R. D., and Segall, A. M. (2004). Unequal access of chromosomal regions to each other in *Salmonella*: Probing chromosome structure with phage λintegrase-mediated long-range rearrangements. *Mol. Microbiol.* **52,** 329–344.

Gordon, G. S., and Wright, A. D. (2000). DNA segregation in bacteria. *Annu. Rev. Microbiol.* **54,** 681–708.

Grindley, N. D. F. (2002). The movement of Tn3-like elements: Transposition and cointegrate resolution. *In* "Mobile DNA II" (N. L. Craig R. Craigie M. Gellert and A. M. Lambowitz, eds.). ASM Press, Washington, DC.

Hecht, R. M., Stimpson, D., and Pettijohn, D. (1977). Sedimentation properties of the bacterial chromosome as an isolated nucleoid and as an unfolded DNA fiber. Chromosomal DNA folding measured by rotor speed effects. *J. Mol. Biol.* **111,** 257–277.

Hendrickson, H., and Lawrence, J. G. (2006). Selection for chromosome architecture in bacteria. *J. Mol. Evol.* **62,** 615–629.

Higgins, N. P. (ed.) (2005). "The Bacterial Chromosome." ASM Press, Washington, DC.

Jayaram, M., Grainge, I., and Tribble, G. (2002). Site-specific recombination by the Flp protein of *Saccharomyces cerevisiae*. *In* "Mobile DNA II" (N. L. Craig R. Craigie M. Gellert and A. M. Lambowitz, eds.). ASM Press, Washington, DC.

Johnson, R. C., Johnson, L. M., Schmidt, J. W., and Gardner, J. F. (2005). Major nucleoid proteins in the structure and function of the *Escherichia coli* chromosome. *In* "The bacterial chromosome" (N. P. Higgins, ed.). ASM Press, Washington DC.

Kavenoff, R., and Bowen, B. C. (1976). Electron microscopy of membrane-free folded chromosomes from *Escherichia coli*. *Chromosoma* **59,** 89–101.

Krug, P. J., Gileski, A. Z., Code, R. J., Torjussen, A., and Schmid, M. B. (1994). Endpoint bias in large Tn10-catalyzed inversions in *Salmonella typhimurium*. *Genetics* **136,** 747–756.

Lanzer, M., and Bujard, H. (1988). Promoters largely determine the efficiency of repressor action. *Proc. Natl. Acad. Sci. USA* **85,** 8973–8977.

Levy, O., Ptacin, J. L., Pease, P. J., Gore, J., Eisen, M. B., Bustamante, C., and Cozzarelli, N. R. (2005). Identification of oligonucleotide sequences that direct the movement of the *Escherichia coli* FtsK translocase. *Proc. Natl. Acad. Sci. USA* **102,** 17618–17623.

Liu, S. L., and Sanderson, K. E. (1996). Highly plastic chromosomal organization in *Salmonella typhi*. *Proc. Natl. Acad. Sci. USA* **93,** 10303–10308.

Manna, D., and Higgins, N. P. (1999). Phage Mu transposition immunity reflects supercoil domain structure of the chromosome. *Mol. Microbiol.* **32,** 595–606.

Manna, D., Breier, A. M., and Higgins, N. P. (2004). Microarray analysis of transposition targets in *Escherichia coli*: The impact of transcription. *Proc. Natl. Acad. Sci. USA* **101,** 9780–9785.

Nash, H. A., and Pollock, T. J. (1983). Site-specific recombination of bacteriophage lambda. The change in topological linking number associated with exchange of DNA strands. *J. Mol. Biol.* **170,** 19–38.

Nielsen, H. J., Jesper, H. J., Ottesen, R., Youngren, B., Austin, S. J., and Hansen, F. G. (2006). The *Escherichia coli* chromosome is organized with the left and right chromosome arms in separate cell halves. *Mol. Microbiol.* **62,** 331–338.

Niki, H., Yamaichi, Y., and Hiraga, S. (2000). Dynamic organization of chromosomal DNA in *Escherichia coli. Genes Dev.* **14,** 212–223.

Palmeros, B., Wild, J., Szybalski, W., Le Borgne, S., Hernandez-Chavez, G., Gosset, G., Valle, F., and Bolivar, F. (2000). A family of removable cassettes designed to obtain antibiotic-resistance-free genomic modifications of *Escherichia coli* and other bacteria. *Gene* **247,** 255–264.

Postow, L., Hardy, C. D., Arsuaga, J., and Cozzarelli, N. R. (2004). Topological domain structure of the *Escherichia coli* chromosome. *Genes Dev.* **18,** 1766–1779.

Rebollo, J. E., Francois, V., and Louarn, J. M. (1988). Detection and possible role of two large nondivisible zones on the *Escherichia coli* chromosome. *Proc. Natl. Acad. Sci. USA* **85,** 9391–9395.

Robinow, C., and Kellenberger, E. (1994). The bacterial nucleoid revisited. *Microbiol. Rev.* **58,** 211–232.

Sambrook, J., Fritsch, E. F., and Maniatis, T. (1989). "Molecular Cloning: A Laboratory Manual," 2nd ed. Cold Spring Harbor Laboratory Press, Cold Spring Harbor, NY.

Schmid, M. B., and Roth, J. R. (1987). Gene location affects expression level in *Salmonella typhimurium. J. Bacteriol.* **169,** 2872–2875.

Segall, A. M., Mahan, M. J., and Roth, J. R. (1988). Rearrangement of the bacterial chromosome: Forbidden inversions. *Science* **241,** 1314–1318.

Sektas, M., and Szybalski, W. (1998). Tightly controlled two-stage expression vectors employing the Flp/FRT-mediated inversion of cloned genes. *Mol. Biotechnol.* **9,** 17–24.

Sektas, M., Hasan, N., and Szybalski, W. (2001). Expression plasmid with a very tight two-step control: Int/att-mediated gene inversion with respect to the stationary promoter. *Gene* **267,** 213–220.

Sinden, R. R., and Pettijohn, D. E. (1981). Chromosomes in living *E. coli* cells are segregated into domains of supercoiling. *Proc. Natl. Acad. Sci. USA* **78,** 224–228.

Ussery, D., Larsen, T. S., Wilkes, K. T., Friis, C., Worning, P., Krogh, A., and Brunak, S. (2001). Genome organization and chromosome structure in *Escherichia coli. Biochimie* **83,** 201–212.

Valens, M., Penaud, S., Rossignol, M., Cornet, F., and Boccard, F. (2004). Macrodomain organization of the *Escherichia coli* chromosome. *EMBO J.* **23,** 4330–4341.

Viollier, P. H., Thanbichler, M., McGrath, P. T., West, L., Meewan, M., McAdams, H. H., and Shapiro, L. (2004). Rapid and sequential movement of individual chromosomal loci to specific subcellular locations during bacterial DNA replication. *Proc. Natl. Acad. Sci. USA* **101,** 9257–9262.

Walhout, A. J., Temple, G. F., Brasch, M. A., Hartley, J. L., Lorson, M. A., van den Heuvel, S., and Vidal, M. (2000). GATEWAY recombinational cloning: Application to the cloning of large numbers of open reading frames or ORFeomes. *Methods Enzymol.* **328,** 575–592.

Wang, X., Liu, X., Possoz, C., and Sherratt, D. J. (2006). The two *Escherichia coli* chromosome arms locate to separate cell halves. *Genes Dev.* **20,** 1727–1731.

Watson, M. A., Boocock, M. R., and Stark, W. M. (1996). Rate and selectivity of synapsis of res recombination sites by Tn3 resolvase. *J. Mol. Biol.* **257,** 317–329.

Wong, M. L., and Medrano, J. F. (2005). Real-time PCR for mRNA quantitation. *BioTechniques* **39,** vol 1. Epub.

[18] Dissecting Nucleic Acid–Protein Interactions Using Challenge Phage

By Stanley R. Maloy and Jeffrey Gardner

Abstract

The bacteriophage P22–based challenge system is a sophisticated genetic tool for the characterization of sequence-specific recognition of DNA and RNA *in vivo*. The construction of challenge phage follows simple phage lysate preparations and detection of constructs by positive selection methods for plaques on selective strains. The challenge phage system is a powerful tool for the characterization of protein–DNA and protein–RNA interactions *in vivo*. The challenge phage has been further developed to characterize the interactions of multiple proteins in heteromultimeric complexes that are required for DNA binding. Under appropriate conditions, expression of the *ant* gene determines the lysis–lysogeny decision of P22. This provides a positive selection for *and* against DNA binding: repression of *ant* can be selected by requiring growth of lysogens, and mutants that cannot repress *ant* can be selected by requiring lytic growth of the phage. Thus, placing *ant* gene expression under the control of a specific DNA–protein interaction provides very strong genetic selections for regulatory mutations in the DNA-binding protein and DNA-binding site that either increase or decrease the apparent strength of a DNA–protein interaction *in vivo*. Furthermore, the challenge phage contains a kanamycin-resistance element that can be used to either directly select for lysogeny or to determine the frequency of lysogeny for a given protein–DNA interaction to measure the efficiency of DNA binding *in vivo*. Selection for lysogeny can be used to isolate DNA-binding proteins with altered or enhanced DNA-binding specificities. The challenge phage selection provides a general method for identifying critical residues involved in DNA–protein interactions. Challenge phage selections have been used to genetically dissect many different prokaryotic and eukaryotic DNA-binding interactions.

Introduction

The bacteriophage-P22 challenge phage system was developed as a genetic approach to characterize the binding of proteins to their specific DNA target sites *in vivo* (Benson *et al.*, 1986). It has been further developed to analyze nucleoprotein complexes. In the bacteriophage lambda integration system, the P22-challenge phage system was used to characterize the

METHODS IN ENZYMOLOGY, VOL. 421   0076-6879/07 $35.00
                    DOI: 10.1016/S0076-6879(06)21018-8

cooperative interactions between the Fis and Xis proteins in DNA sequence-specific binding (Numrych *et al.*, 1991). In the Hin recombinase system that mediates a site-specific recombination in the *Salmonella* chromosome, the challenge phage system was used to dissect Hin mutants specific to different steps in the recombination reaction including dimer formation, tetramer formation, interactions with Fis protein and DNA cleavage (Nanassy and Hughes, 1998). The system has also been adapted to study RNA–protein interactions *in vivo* (Fouts and Celander, 1996; MacWilliams *et al.*, 1993; Wang *et al.*, 1997). The fundamental utility of this system is that it provides positive selections for mutants that either lose binding to specific DNA or RNA sequences or gain the ability to bind altered DNA or RNA target sequences. If the sequence-specific binding to DNA or RNA requires the formation of multiple protein complexes, the challenge phage system can be used to genetically dissect these complexes. The principle behind the method is to make the P22 lysis/lysogeny decision dependent on the binding of specific proteins to their specific DNA or RNA target sites. This is done by placing a phage reporter gene under control of either the specific DNA sequence or the DNA sequence coding for the specific RNA sequence to be characterized (Bass *et al.*, 1987; Maloy and Youderian, 1994). This approach can be applied to a wide variety of DNA-binding proteins (Table I) because most sequence-specific, DNA-binding proteins can act as repressors if their binding sites are placed close enough to the promoter of the reporter gene in the P22 challenge phage system (Gralla and Collado-Vides, 1996).

Regulation of Lysis and Lysogeny P22

The P22 challenge phage assay takes advantage of the regulatory properties of the immunity I region of the *Salmonella*-specific bacteriophage P22 (Susskind and Youderian, 1983). Bacteriophage P22 is related to the bacteriophage lambda, but unlike phage lambda, P22 carries two immunity regions that control the lysis/lysogeny decision (Susskind and Botstein, 1978). Control of the lysis/lysogeny decision in bacteriophage lambda is under the cI repressor binding to the operator sites at the P_L and P_R promoters within the immunity C region. Bacteriophage P22 has a similar immunity C region, but the lysogenic repressor is called c2. The role of the immunity I region in bacteriophage P22 is to control expression of the *ant* gene. The P22 *ant* gene encodes antirepressor, or Ant. Ant binds directly to the c2 lysogenic repressor of P22 to inhibit DNA binding and prevent P22 lysogeny. When wild-type P22 is plated on a lawn of sensitive *Salmonella* cells, roughly half of the phage lysogenize and the other half grow lytically to yield about 1000 phage particles per single infected cell. The phage

TABLE I

EXAMPLES OF DNA–PROTEIN AND RNA–PROTEIN INTERACTIONS THAT HAVE CHARACTERIZED USE OF CHALLENGE PHAGE[a]

	Source	References
DNA–protein interaction		
λ repressor	Phage λ	Benson and Youderian, 1989
Cro repressor	Phage λ	Benson and Youderian, 1989
Integrase	Phage λ	Han et al., 1994; Lee et al., 1990
Fis-Xis-Int	Phage λ	Numrych et al., 1991
attL complex	Phage λ	MacWilliams et al., 1997
EcoRI	RY13 plasmid	Szegedi and Gumport, 2000
TetA repressor	Tn10	Benson et al., 1986
TrpR repressor	E. coli	Bass et al., 1987
LacI repressor	E. coli	Benson et al., 1986
GalR repressor	E. coli	Benson et al., 1986
Integration host factor	E. coli	Hales et al., 1994; Lee et al., 1992
Hin recombinase	S. typhimurium	Hughes et al., 1988
Hin-Fis interactions	S. typhimurium	Nanassy and Hughes, 1998
PutA repressor	S. typhimurium	P. Ostrovsky and S. Maloy, unpublished
Nac activator	Klebsiella aerogenes	Chen et al., 1998
OxyR/MtrA	Mycobacterium	T. Zahrt and V. Deretic, unpublished
nifH promoter	Rhizobium meliloti	Ashraf et al., 1997
ToxR	Vibrio cholerae	Pfau and Taylor, 1998
FLP recombinase	S. cervesiae	Lebreton et al., 1988
Estrogen Response Element	H. sapiens	Chusacultanachai et al., 1999
RNA–protein interaction		
Coat proteins	Phage R17 and MS2	Celander et al., 2000

[a] Numerous other DNA–protein interactions have been studied as well. The critical requirement for any specific DNA–protein interaction is that the protein is expressed and active in Salmonella.

released after lytic infection goes on to infect surrounding cells where the lysis/lysogeny decision is repeated. Again roughly half lysogenize and half grow lytically. The end result is the formation of a turbid plaque on a lawn of sensitive cells. The turbidity is the result of growth of lysogenic cells that are "immune" to further infection by P22. If Ant is expressed constitutively, then P22 growth is lytic; the phages do not lysogenize. The constitutive lytic growth of P22 is directly visualized on agar media by the formation of clear plaques when the phages are plated on a lawn of sensitive *Salmonella* cells. If the *ant* gene is deleted, the phages form turbid plaques similar to wild-type P22. The purpose of Ant is believed to allow growth on cells lysogenic for P22-related phages.

Regulation of *ant* Gene Expression

The *ant* gene of bacteriophage P22 is controlled by at least three mechanisms (Susskind and Youderian, 1983). The *mnt* gene (*maint*enance of lysogeny) encodes a repressor that binds a DNA sequence at the *ant* promoter. Mnt will inhibit *ant* gene expression in both cis and trans. Once lysogenized, Mnt produced from the P22 lysogen will prevent expression of *ant* from further P22 phage infection. The products of two other genes, *arc* and *sar*, modulate the level of expression of the *ant* gene promoter. The *arc* gene is transcribed in an operon with *ant* (*arc-ant*) while the *sar* gene is within the *arc-ant* operon but expressed from the noncoding strand. The *sar* gene encodes an antisense RNA that inhibits translation of the *arc-ant* transcript. The *mnt* and *arc-ant* operon promoters are divergently transcribed, and Arc binds to sequences between these promoters and represses transcription from both. Thus, the *ant* gene is highly controlled for reasons that are not completely understood to carefully control the level of Ant and possibly allow the phage to sense the metabolic state of the host cell in making the lysis/lysogeny decision.

Selection For and Against DNA Binding in the
Challenge Phage System

The challenge phage system uses the P22 *ant* gene as a reporter for repression at artificial DNA operator sites that replace the wild-type P*ant* operator sequence that interacts with Mnt. The phages are engineered so that expression of *ant* determines the lysis–lysogeny decision of P22 (Benson *et al.*, 1986). A constitutively expressed kanamycin-resistance gene replaces the *mnt* gene in the challenge phage. The Arc repressor is removed in the challenge phage by the introduction of an amber mutation in the *arc* gene. The absence of Mnt and Arc results in constitutive expression of Ant.

This corresponds to lytic growth unless P*ant* can be repressed by the binding of a protein to the artificial P*ant* operator site. Thus, expression of *ant* is controlled by a specific DNA interaction to provide a simple, sensitive genetic selection for mutations in the DNA-binding site that increase or decrease the strength of the DNA–protein interaction. In addition, binding at P*ant* can be directly measured because lysogeny of the challenge phage confers kanamycin resistance (Km^R) to the cell. These features provide general methods for measuring relative *in vivo* affinities of DNA-binding proteins for target sites and for identifying specific base pairs in DNA that contact the protein. Mutants in the DNA target sequence of a challenge phage that are defective in binding to the target protein are selected for directly by plating the challenge phage on cells that express the protein that binds the DNA target. Phages with altered DNA sites defective in interaction with the target protein form plaques, while the parent challenge phage does not. Thus, the challenge phage provides a simple positive selection for DNA target site mutants. In addition, the frequency of lysogeny of these mutant phages, as measured by the frequency of Km^R lysogens per infected cell, is an indirect measure of the degree of the defect in DNA interaction to a specific DNA mutant site.

In addition to a positive selection for DNA target site mutants defective in binding, the challenge phage can also be used to select for protein mutants with altered or enhanced binding specificities. Once a specific base pair within a DNA target site is identified as important for direct interaction with its cognate protein, a challenge phage carrying the mutant target site can be used to find mutant proteins that acquire the ability to bind the mutant site in the challenge phage. A defect in binding results in lytic growth and the inability to lysogenize. Because lytic growth kills the cell, the frequency of Km^R lysogens is reduced up to 10^8-fold. This provides a powerful selection for altered proteins that bind the mutant site to repress at P*ant* and allow lysogeny and growth in the presence of kanamycin. Proteins that bind mutant DNA sites can have either altered or enhanced DNA-binding activity. Mutants with altered binding activity have the ability to bind the mutant site, but lose the ability to bind the wild-type sequence. Mutants with enhanced binding activity are able to bind both mutant and wild-type sites.

Construction of Challenge Phage with Novel Operator Sequences at P*ant*

Challenge phages are constructed in two steps. First, the DNA-binding site is cloned into a plasmid that carries the P22 *immI* region. The plasmid pPY190 is a pBR322 derivative that carries a 500 bp *Eco*RI-*Hin*dIII fragment from the *immI* region (Fig. 1). The *immI* DNA cloned onto pPY190 includes

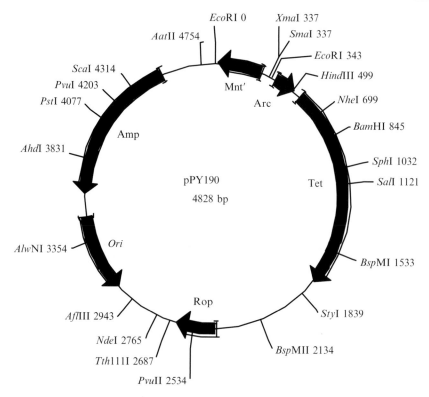

FIG. 1. Plasmid pPY190. pPY190 is a 4.8-kb plasmid with the 5′ end of the *mnt* gene and the entire *arc*+ gene from the P22 *immI* region cloned into the *Eco*RI-*Hin*dIII sites of pBR322. The O*mnt* was replaced with a *Sma*I/*Xma*I substitution constructed by site-directed mutagenesis. Cloning at *Sma*I places the insert at –3, and cloning at *Xma*I places the insert at –1 relative to the transcription start site of *ant*.

P*ant* and the 5′ end of *arc*. The O*mnt* site has been replaced with *Sma*I and *Eco*RI sites. Blunt-ended DNA fragments such as complementary oligonucleotides or restriction fragments can be cloned into the *Sma*I site. This places the insert 3 bp upstream of the initiation site of P*ant*. The presence of the insert can be easily verified by PCR analysis.

It is also useful to clone a DNA-binding site of interest with an additional two and four bases flanking the binding sequence. Since there are about 10 base pairs per turn of the DNA helix, adding two and four bases with rotation by 72° and 144° where the new repressor will bind relative to where RNA polymerase will bind at P*ant*, may improve repression by the protein of interest.

P22-1000 is a backbone phage with the *mnt* gene substituted by a kanamycin-resistance gene and an amber mutation in the *arc* gene. The substitution in pPY190 can be moved onto phage P22-1000 (P22 *mnt*::Km *arc*^{Am}) by homologous recombination *in vivo*. A host containing pPY190 with the novel DNA-binding site cloned into the P*ant* operator site is infected with the phage and a lysate is prepared. During lytic growth, the phage and the plasmid can undergo recombination between the sequence homology on the plasmid and the phage. Recombinant phages that contain the substituted DNA-binding site in place of O*mnt* can be identified by infecting strain MS1582. MS1582 contains a P22 prophage that expresses the c2 and Mnt proteins. Mnt represses O*mnt* of incoming P22-1000 parental phages preventing lytic growth and do not form plaques on MS1582. In contrast, recombinant phages carrying the substituted DNA-binding site are not repressed by Mnt, and thus express Ant, forming plaques. One caveat is that if the *Salmonella* host chromosome encodes the protein that recognizes the DNA-binding site, it should be mutated so that the protein does not repress at P*ant* in the new challenge phage construct.

Since the parental phage carries the *arc*^{Am} mutation and the pPY190 plasmid contains the wild-type allele, recombinants are either *arc*^{Am} or *arc*$^+$, depending on which side of the mutation the crossover occurred. Both types of recombinants form plaques on MS1582. The *arc*$^+$ recombinants form small, turbid plaques because Arc can partially repress P*ant*. The *arc*^{Am} recombinants form large, clear plaques that can be purified and used in subsequent experiments.

The desired recombinants contain both the DNA-binding site and the *arc*^{Am} mutation. The O*mnt* substitution in pPY190 contains an *Eco*RI site that is not present in the parental phage, and the *arc*^{Am} mutation destroys a *Fnu*4H site (Fig. 2). Thus, the desired recombinants can be confirmed by PCR amplification of the *immI* region followed by restriction fragment–length polymorphism (RFLP) mapping with *Eco*RI and *Fnu*4H (Fig. 3).

Essentially, any DNA-binding site can be cloned into pPY190 and recombined onto P22 to form a challenge phage. It is also possible to clone large fragments that contain multiple DNA-binding sites. For simple DNA–protein interactions, the extent of repression of P*ant* by a DNA-binding protein depends on the proximity of the protein to the transcription start site. For example, a consensus λ-cI binding site functions well when placed at the +1 or +4 position, but does promote repression when placed at position +24 (Gralla and Collado-Vides, 1996). It is also possible to construct phages where DNA looping regulates expression from P*ant*. In the case of the λ *attL* site, repression required an IHF-mediated loop containing an Int protein monomer bound simultaneously to both the P*ant* region and a sequence more than 50 bp downstream of P*ant* (MacWilliams *et al.*, 1997).

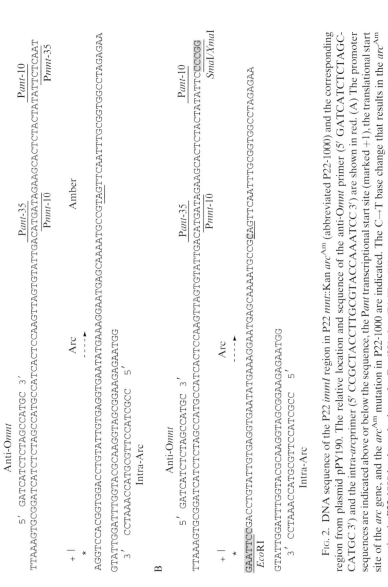

Fig. 2. DNA sequence of the P22 *immI* region in P22 *mnt*::Kan *arc*^{Am} (abbreviated P22-1000) and the corresponding region from plasmid pPY190. The relative location and sequence of the anti-O*mnt* primer (5′ GATCATCTCTAGC-CATGC 3′) and the intra-*arc*primer (5′ CCGCTACCTTGCGTACCAAATCC 3′) are shown in red. (A) The promoter sequences are indicated above or below the sequence, the P*ant* transcriptional start site (marked +1), the translational start site of the *arc* gene, and the *arc*^{Am} mutation in P22-1000 are indicated. The C→T base change that results in the *arc*^{Am} mutation in P22-1000 leads to the loss of a *Fnu*4H1 site. (B) The differences in the pPY190 sequence are highlighted in yellow. A substitution places a *Sma*I/*Xma*I and *Eco*R1 site adjacent to the +1 site of the *arc* gene. In addition, pPY190 has a *Fnu*4H1 site at the position corresponding to the *arc*^{Am} mutation in P22-1000.

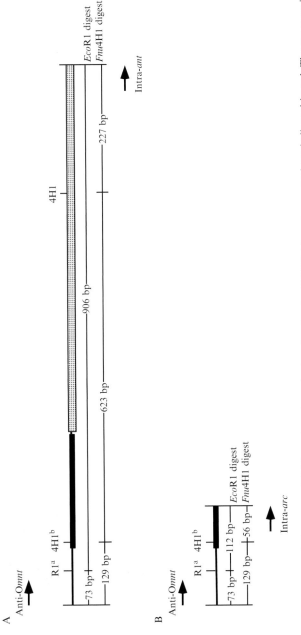

FIG. 3. Restriction maps of PCR products from the *immI* region in pPY190. The primers used are indicated in red. The *arc* gene is indicated by the black box, and the *ant* gene is indicated by the gray box. The *Eco*R1 site labeled R1[a] and the *Fnu*4H1 site labeled 4H1[b] are present in pPY190 but not in P22-1000. The presence of a cloned fragment in the *Sma*I or *Xmn*I site will increase the size of the 73 bp *Eco*R1 and 129 bp *Fnu*4H1 fragments. (A) PCR with the anti-O*mnt* and Intra-*ant* primers yields a 979-bp fragment. (B) PCR with the anti-O*mnt* and intra-*arc* primers yields a 185-bp fragment.

Challenge phages containing *ant::lacZ* or *ant::cam* fusions can also be constructed by genetic recombination. Isolation of the *ant::lacZ* or ant::cam lysogens provides a single-copy chromosomal fusion as a reporter gene (Hughes *et al.*, 1988).

Experimental Procedures

Procedure 1. Growing P22 Lysates from Single Plaques

Media and Reagents

Luria broth (LB): 10 g tryptone, 5 g yeast extract, 5 g NaCl per liter of deionized water

LB phage plates: LB + 12 g/l Bacto Agar per liter of deionized water

Top agar: 10 g tryptone, 7 g agar per liter of deionized water

Sterile saline: 8.5 g NaCl per liter of deionized water

BS (buffered saline): 8.5 g NaCl per liter of deionized water

E medium (Maloy *et al.*, 1996; Vogel and Bonner, 1956): 5× E medium contains 1 g $MgSO_4 \cdot 7H_2O$, 10 g citric acid $\cdot H_2O$, 50 g K_2HPO_4, and 17.5 g $NaHNH_4PO_4 \cdot 2H_2O$ in 1000 ml of deionized H_2O

LBEG (Maloy *et al.*, 1996): LB 0.2% D-glucose 0.5x E-salts

TBSA (Maloy *et al.*, 1996): 10 g tryptone, 5 g NaCl, and 7 g agar in 1000 ml of deionized H_2O

TE: 10 m*M* TrisHCl, 1 m*M* EDTA, pH 8

Antibiotics

When needed, add antibiotics to LB medium at the following concentrations: 50 µg/ml kanamycin sulfate (Km), 15 µg/ml tetracycline hydrochloride (Tc), and 100 µg/ml sodium ampicillin (Ap).

Bacteria and Phage

A typical P22 lysate contains about 10^{11} phage per ml, and is stable for many years when stored over chloroform at 4°. Lytic growth is favored when the host cells are growing rapidly and the multiplicity of infection (MOI) is low. Lysogeny is favored when cells are growing slowly and the MOI is high. When the MOI of P22 is greater than 10, more than 95% of the infected cells form lysogens (Levine, 1957). P22 diffuses rapidly throughout a soft agar overlay on a petri dish upon extended storage. This can result in contamination of one plaque with phage from a nearby plaque. Therefore, to ensure the genetic homogeneity of phage lysates, recombinant phage should be purified by streaking for isolated, single plaques several times before growing large-scale lysates.

Strains

Strains used in these protocols are listed in Table II. Although *S. enterica* strains can cause gastrointestinal infections in humans, the *S. enterica* serovar Typhimurium (formerly *S. typhimurium*) strain LT2 strains used for challenge phage assays are attenuated for virulence (Sanderson and Hartman, 1978; Slauch *et al.*, 1997). However, *S. typhimurium* can cause infections in immunocompromised hosts and many of the strains carry drug-resistance plasmids or transposons that could be transferred to other enteric bacteria. Therefore, good microbiological technique should be followed when working with *Salmonella* (see the Centers for Disease Control and Prevention recommendations for biosafety in microbiological and biomedical laboratories at http://siri.org/library/cdc.html).

STANDARD P22 LYSATE PREPARATION FROM SINGLE PLAQUES. Grow the recipient P22-sensitive strain overnight in LB. Add a single plaque by inserting a Pasteur pipette into a plate that contains plaques of the phage to be grown, twisting the pipette to separate the agar and transfer the plaque into the overnight recipient culture by simply blowing the plaque into the culture. Add 4 ml of LBEG and grow an additional 6 hr with shaking (aeration) at 37°. Pellet cells and pour lysate into sterile tube, add a few drops of $CHCl_3$, vortex, and store at 4°. Titer the lysate by mixing 0.1 ml of a serial dilution with 0.1 ml of an overnight culture of sensitive cells, add 3 ml of molten top agar cooled to ~50° (i.e., you can hold it in your hand and it does not hurt), and pour directly into a LB phage plate, moving the plate so the molten agar covers the plate surface. It usually takes 6 h at 37° for plaques to become visible. Most P22 titers are 10^{10} to 10^{11} pfu (plaque-forming units) per milliliter.

STANDARD HIGH-TITER P22 LYSATE PREPARATION FROM SINGLE PLAQUES. For high-titer phage stocks (10^{12} pfu/ml), subculture the recipient cells by diluting 1/10 into 30 ml of fresh LBEG medium and grow to an OD_{600} of ~0.8. Add a single plaque (Pasteur pipette plug) and grow at 37° until visible lysis occurs. For P22 phages with the arc^{Am} mutation, growth at 30° is optimal. Lysis occurs after about 4 h, but overnight growth is fine. Add 2 ml of $CHCl_3$ and shake an additional 5 min. Transfer aqueous layer to a 40-ml Oakridge centrifuge tube. Pellet debris by centrifugation in a SS34 rotor at 5000 rpm for 5 min and transfer the supernatant to a new Oakridge centrifuge tube. Spin in a SS34 rotor at 16,000 rpm for 1 h to pellet phage. Pour off the supernatant and add 4 ml of BS (buffered saline). Let sit 2 h at room temperature to allow pellet to loosen. Resuspend phage by vortexing

TABLE II
HOST STRAINS FOR CHALLENGE PHAGE EXPERIMENTS

Strain	Species[a]	Genotype	Source
MS1363	St	leuA414(Am) Fels⁻ supE40	M. Susskind[b]
MS1868	St	leuA414(Am) Fels⁻ hsdSB (r⁻ m⁺)	M. Susskind
MS1882	St	leuA414(Am) Fels⁻ hsdSB (r⁻ m⁺) endA	M. Susskind
MS1883	St	leuA414(Am) Fels⁻ hsdSB (r⁻ m⁺) supE40	M. Susskind
MS1582	St	leuA414(Am) Fels⁻ supE40 ataA::[P22 sieA44 16(Am)H1455 Tpfr49]	Benson et al., 1986
MST2781	St	leuA414(Am) Fels⁻ hsdSB (r⁻ m⁺) supE40 ataA::[P22 sieA44 Δ(mnt-al Tn1) Amps 9⁻ Tpfr184]/pGW1700 (mucA⁺ B⁺ Tet⁺)	Maloy et al., 1996
MST2786	St	leuA414(Am) Fels⁻ hsdSB (r⁻ m⁺) / pGW1700 (mucA⁺ B⁺ Tet⁺)	Maloy et al., 1996
TH564	St	leuA414(Am) Fels⁻ supE40 ataA::[P22 sieA44 Δ(mnt-al Tn1) Amps 9⁻ Tpfr184]	Hughes et al., 1988
TH1901	St	leuA414(Am) Fels⁻ hsdSB (r⁻ m⁺) supE40 ataA::[P22 Ap521]	Maloy et al., 1996
EM425	Ec	F⁻ supE44 recA1 gyr-96 thi⁻1 endA hsdR17 (r⁻ m⁻) λ⁻ / pPY190	

[a] St indicates Salmonella enterica serovar Typhimurium LT2 derivatives, and Ec indicates Escherichia coli K-12 derivatives. A kit with these strains can be obtained from the authors.

[b] Dr. Mimi Susskind, Department of Molecular Biology, University of Southern California, Los Angeles, CA.

and transfer to a sterile tube with ChCl$_3$. This provides a high-titer lysate ($\sim 10^{12}$ pfu/ml) from which DNA is readily extracted.

Protocols for construction and use of challenge phage are detailed below. Basic phage biology, molecular biology, and genetic protocols can be found in Maloy *et al.* (1996).

Procedure 2. Construction of Challenge Phage with Novel Operator Sites at *P*ant

Materials

Strain MS1883 carrying a pPY190 derivative with DNA-binding site cloned into the *Sma*I site.

Construction of Challenge Phage by In Vivo *Recombination*

1. Grow the MS1883 containing a pPY190 subclone with the desired DNA-binding site in LB plus Ap at 37° overnight.
2. Mix 10^6 to 10^7 plaque-forming units (pfu) of P22-1000 (*mnt*::Km-9 *arc*^{Am}) phage with 50 μl of the overnight culture. Leave at room temperature for 5 min to allow phage adsorption.
3. Add 2 ml LBEG and grow at 37° with vigorous aeration for 3 h or until the culture is lysed.
4. Add several drops of CHCl$_3$ and vortex thoroughly. Allow the CHCl$_3$ to settle. Pour the supernatant (avoiding the CHCl$_3$) into a microfuge tube. Centrifuge 2 min in a microfuge to pellet the cell debris.
5. Transfer the supernatant to a new tube, add a few drops of chloroform, vortex, and allow the chloroform to settle. Store the lysate at 4°.
6. Dilute the lysate 1/100 by adding 10 μl to 1 ml LB. Vortex. Mix 0.2 ml of the diluted lysate with 0.1 ml of a fresh overnight culture of MS1582 (to give a MOI < 1).
7. Leave at room temperature for 5 min to allow phage adsorption.
8. Add 3 ml of melted top agar to a test tube, place in a 50° heating block, and allow the temperature of the top agar to equilibrate to 50°.
9. Add 3 ml molten, 50° top agar to the cells plus phage, mix, and quickly pour onto a LB-phage plate. Allow the top agar to solidify at room temperature, and then incubate overnight at 37°.
10. Pick large clear plaques with sterile toothpicks and streak for single plaques on a LB-phage plate overlaid with 3 ml top agar containing 0.1 ml of fresh MS1582. Incubate overnight at 37°. This is best done by cooling the plates after the addition of top agar–containing cells

at 4° for 10 min to solidify the top agar, followed by streaking with sterile toothpicks or a cooled, flame-sterilized 32-gage platinum wire.

11. Pick large, clear, well-isolated plaques, and streak for isolation on a LB-phage plate overlayed with 3 ml top agar containing 0.1 ml of fresh MS1883. Incubate overnight at 37°.

12. Pick and re-streak the large, clear plaques. Incubate overnight at 37°. The larger, clear plaques contain potential recombinant challenge phage.

Procedure 3. Identification of Challenge Phage Recombinants by RFLP Mapping

The desired challenge phage recombinants have a restriction fragment length polymorphism (RFLP) that can be identified by digesting the *immI* PCR fragment with *Eco*R1 or *Fnu*4H1.

Preparation of Challenge Phage Lysates

1. For each potential recombinant challenge phage to be tested, add 0.1 ml of MS1883 to 5 ml LBEG. Prepare an identical culture as a control. Incubate on a shaker at 37° until the cells reach early exponential phase (about 1 to 2 h).

2. Using a sterile Pasteur pipette, plug an independent, well-isolated, large clear plaque from the potential challenge phage, and add one plaque to each tube of MS1883. To the control culture, add a plaque of the P22-1000 parental phage. Incubate at 37° with vigorous aeration for about 3 h or until lysed.

3. Add several drops of $CHCl_3$, vortex, and allow the chloroform to settle. Transfer the supernatant (avoiding the $CHCl_3$) to microfuge tubes. Centrifuge for 2 min in a microfuge to pellet the cell debris.

4. Transfer the supernatant from each phage into clean microfuge tubes. Centrifuge for 30 min at 4° in a microfuge to pellet the phage.

5. Carefully pour off the supernatant. Add 0.1 ml 0.85% NaCl to each phage pellet and vortex to resuspend the phage. Combine the supernatants from a single phage into a single tube. Add several drops of chloroform and vortex. Store the concentrated lysate at 4°.

PCR and RFLP Analysis

1. Phage lysates can be amplified directly without purifying phage DNA. Include the parent P22-1000 and a challenge phage with the P*ant* region of pPY190 recombined onto P22 *arc*^{Am} as controls.

2. Combine the following in a 500 μl microfuge tube:
   a. 10 μl P22 lysate
   b. 1 μl intra-Arc primer (20 pmol; 5′ CCG CTA CCT TGC GTA CCA AAT CC 3′) *or* intra-Ant primer (20 pmol; 5′ CAA GGC TGT TTG CTT CTT TTG CAG 3′)
   c. 1 μl anti-O*mnt* primer (20 pmol; 5′ GAT CAT CTC TAG CCA TGC 3′)
   d. 37 μl sterile deionized H₂O
   e. 10 μl 10× PCR buffer (supplied by the manufacturer)
   f. 10 μl each of 2 m*M* dATP, dGTP, dTTP, and dCTP
   g. 1 μl Taq DNA polymerase (Perkin-Elmer Corporation, 2.5 units/μl)
3. Mix and amplify in a thermocycler using the following program:
   Step 1, 95°, 90 sec
   Step 2, 53°, 30 sec
   Step 3, 74°, 60 sec
   Step 4, cycle steps 1 to 3, repeat 30 times
   Step 5, 74°, 5 min
   Step 6, 4°, hold
4. Purify the PCR products by phenol extraction followed by ethanol precipitation or by using a commercially available method such as Qiagen Affinity columns (Qiagen), Wizard Preps (Promega), or Millipore MC filters (Millipore).
5. In separate tubes, digest 10 μl of the PCR product with *Eco*RI and *Fnu*4H (NEB).
6. Electrophorese the DNA and undigested controls on a 3.0% NuSieve GTG (FMC) or 3:1 (Sigma) agarose gel in 1X TBE at 100 volts until the bromphenol blue is 3/4 the length of the gel. Include a lane with low-molecular-weight standards.
7. Stain the gel with ethidium bromide. Visualize the gel with UV and photograph the gel. Compare the restriction digests with the expected results for P22-1000 and pPY190 shown in Fig. 3.

Procedure 4. Challenge Phage DNA-Binding Assays

The challenge phage assay depends on binding of a protein to the cognate DNA-binding site at P*ant*. To vary the levels of a DNA-binding protein for use with challenge phage, it is usually necessary to express the DNA-binding protein under control of a regulated promoter, such as P*lac*, P*tac*, or P*bad*. The genetic background of the host strain will depend on the type of regulated promoter used: for P*lac* and P*tac*, the *lacI*^{Q} gene should be provided on a plasmid, and for P*bad*, the chromosomal copy of the *ara* operon should be deleted.

When a DNA-binding protein is expressed from a regulated promoter, the relative *in vivo* affinity of a DNA-binding protein for the binding site on the challenge phage can be quantitated by measuring the frequency of Km^R lysogens in media with different concentrations of inducer (and hence different concentrations of the DNA-binding protein). The relative DNA-binding affinity can be expressed by plotting the log (percent lysogeny) versus the log (inducer).

The procotols below are based on the regulated expression of a DNA-binding protein from the P*tac* promoter under the control of the *lacI*^{Q} gene product. In this case, expression of the DNA-binding protein is induced by the addition of isopropyl-β-D-galactopyranoside (IPTG). Typically, 0.1 to 1 m*M* IPTG results in maximal induction from P*tac*. However, the optimal concentration of IPTG for each specific DNA-binding protein should be determined empirically by testing 10-fold dilutions of IPTG to ascertain the concentration that promotes maximal lysogeny of the challenge phage. The cell viability should be measured at each concentration of inducer because overexpression of DNA-binding proteins often inhibits cell growth. All of the dilutions can be done in microtiter plates. Using a multichannel pipettor, 48 samples can be spotted onto each agar plate for phenotypic selection or screening.

Challenge Phage Assays

1. Titer the challenge phage on strain MS1883.

2. Subculture 50 μl of MS1868 cells with the P*tac*-DNA–binding protein expression plasmid into 2 ml LB containing an appropriate antibiotic to maintain selection for the expression plasmid. Incubate on a shaker at 37° for about 1 to 2 h until the cells reach early exponential phase.

3. Subculture 0.5 ml of the early exponential phase culture into 1.5 ml of LB + antibiotic, and LB + antibiotic + IPTG. Incubate on a shaker at 37° for about 1 to 2 h until the cells reach early exponential phase.

4. Prepare serial 10-fold dilutions of both cultures to 10^{-5} in LB. To determine the viable cell count, remove 5 μl from the 10^{-3}, 10^{-4}, and 10^{-5} dilutions and spot onto LB-agar plates. Add 100 μl LB to each spot and spread. Incubate the plates overnight at 37°.

5. Calculate the volume of each phage required to give multiplicity of infection (MOI) of 10 to 50 pfu/cell. For example, if the cultures are at about 10^8 cells per milliliter, for an MOI of 25, add the calculated volume of each challenge phage lysate to 0.1 ml of the undiluted cells. Include a P22-1000 phage control.

6. Mix the cells and phage by gently pipetting the solution up and down, and then leave at room temperature for 1 h to allow phage adsorption and phenotypic expression.

7. Prepare sequential 10-fold dilutions of each infected cell culture to 10^{-4} in LB.

8. Remove 10 μl from the 10^{-1}, 10^{-2}, 10^{-3}, and 10^{-4} dilutions of infected cells grown without IPTG and spot onto a LB-agar-Kan + antibiotic plate.

9. Remove 10 μl from the 10^{-1}, 10^{-2}, and 10^{-3} dilutions of the cells grown with IPTG and spot onto a LB-agar-Km + antibiotic + IPTG plate.

10. Allow the spots to dry at room temperature, and then incubate 1 to 2 days at 37°.

11. Count the number of colonies on each plate. Calculate the number of Km^R lysogens per viable cell. If the DNA-binding protein is properly expressed, challenge phage will form Km^R lysogens at a much higher frequency in cells grown with inducer than in cells grown without inducer, and the control phage will not form Km^R lysogens at a significant frequency in either condition.

Procedure 5. Selection for DNA-Binding Site Mutations

Challenge phage form lysogens on a host that expresses a DNA-binding protein that binds to the substituted O_{mnt} site. It is possible to select for "operator constitutive" (O^c) mutations in the DNA-binding site by isolating challenge phage mutants that are not repressed by the DNA-binding protein and thus form plaques on a lawn of cells that express the DNA-binding protein. This provides a very strong selection for mutations that affect the DNA–protein interaction *in vivo*. Mutations in a DNA-binding site are usually rare, but the resulting rare plaques can be easily selected from a population of more than 10^8 infecting phages.

DNA-binding site mutants may result from spontaneous mutagenesis. However, spontaneous mutants are often due to deletion mutations that are less valuable for defining DNA–protein interactions than point mutations. The frequency of point mutations may be enhanced by mutagenizing the phage. It may be useful to use several mutagens with different specificities to obtain a wide spectrum of mutations (Maloy *et al.*, 1996). Two methods of mutagenesis that are particularly useful are UV mutagenesis and hydroxylamine mutagenesis. Phages can be directly treated with both of these mutagens *in vitro*. UV-enhanced base substitution is greatly enhanced by exposing the challenge phage to UV and selecting in a strain expressing the DNA-binding protein that also carries plasmid pGW1700 (Tc^R $mucA^+B^+$) or plasmid pGW249(Km^R $mucA^+B^+$). Plasmids pGW1700 and pGW249 express DNA-polymerase accessory repair proteins that allow for a high frequency of base substitution mutations during the repair of UV-damaged DNA.

Using this approach, it is possible to quickly isolate a large number of O^c mutations in the substituted DNA-binding site. The resulting mutations define critical residues of the DNA-binding site necessary for recognition by a DNA-binding protein. Many DNA-binding sites are palindromes bound by identical subunits of dimeric regulatory proteins: Each monomer recognizes one of the half-sites. For palindromic DNA-binding sites, it may be necessary to construct symmetric double mutants to determine their effect on DNA binding. Such symmetric double mutants can be constructed by cloning the appropriate double-stranded oligonucleotide into the *Sma*I site of pPY190 to yield the corresponding challenge phage.

In the following procedure, the challenge phages are mutagenized with UV light. An *S. typhimurium* strain carrying a plasmid that expresses the cognate DNA-binding protein, and a second compatible plasmid carrying the $mucA^+B^+$ genes is infected with the irradiated phage. Clear plaque mutations that inactivate the binding site will be rare (typically about 10^{-6} to 10^{-7}), and the background will have numerous faint turbid plaques. Therefore, it is necessary to plaque purify the clear plaque mutants several times.

Isolation of Mutations in a DNA-Binding Site

1. Grow an overnight culture of the MST2781 cells with the P_{tac}-DNA–binding protein expression plasmid in 2 ml LB + Tet + antibiotic at 37°.
2. Subculture 0.1 ml of the overnight culture into 2 ml LB + Tet + antibiotic. Incubate with vigorous aeration at 37° for 2 to 3 h or until mid-exponential phase.
3. Dilute 0.5 ml of the mid-exponential phase culture into 1.5 ml LB + Tet + antibiotic + IPTG. Incubate the cells for 1 to 2 h with vigorous aeration at 37°.
4. Dilute the challenge phage to approximately 10^9 pfu/ml in 0.85% NaCl (final volume needed is 1 ml). Save 0.5 ml of the diluted phage as a control.
5. Place 0.5 ml of the diluted phage in a small sterile petri dish, remove the lid, and UV irradiate to 12,000 μJ/cm^2.
6. Prepare serial 10^{-1} and 10^{-2} dilutions of the UV-irradiated phages and the control phages in 0.85% NaCl.
7. Mix 200 μl of the IPTG-induced culture with 100 μl of each phage dilution.
8. Leave at room temperature for about 20 min to allow phage adsorption.

9. To each mixture of cells + phage, add 3 ml TBSA top agar that has been melted and then cooled to 50°. Mix gently, and then quickly pour onto an LB-agar plate and swirl the plate to cover the surface evenly. Allow the top agar to solidify for about 15 min at room temperature, and then incubate overnight at 37°.
10. Examine the plates for clear plaques.

Purify Clear Plaque Mutants

1. Subculture 0.1 ml of the overnight culture of TH564 cells with the P_{tac}-DNA–binding protein expression plasmid into 2 ml LB + antibiotic. Incubate with vigorous aeration at 37° about 2 to 3 h until the cells reach mid-exponential phase.
2. Dilute 0.5 ml of the mid-exponential phase culture into 1.5 ml LB + antibiotic + 1 mM IPTG. Incubate the cells for 1 to 2 h with vigorous aeration at 37°.
3. Add 0.1 ml of the culture induced in LB + antibiotic + 1 mM IPTG to 3 ml TBSA that have been melted and then cooled to 50°. Mix gently and then quickly pour onto an LB + 1 mM IPTG plate and swirl the plate to cover the surface evenly. Allow the top agar to solidify at room temperature.
4. Pick large clear plaques from challenge phages with an O^c mutation in the DNA-binding site and streak on the cell lawn. Also streak out the unmutagenized challenge phage and P22-1000 as controls. Incubate the plates at 37° overnight.
5. The next day, add 0.1 ml of a fresh, overnight culture of MS1883 to 3 ml TBSA that have been melted and cooled to 50°. Mix gently and then quickly pour onto an LB plate and swirl the plate to cover the surface evenly. Allow the top agar to solidify at room temperature.
6. Pick large clear plaques from step 4 and re-streak on the MS1883 lawn. Incubate the plates at 37° overnight.
7. Remove a well-isolated, large clear plaque from each streak with a sterile Pasteur pipette.
8. Add the agar plug to 2 ml LBEG. Vortex. Add 0.1 ml of an overnight culture of MS1883. Incubate with vigorous aeration at 37° for about 3 h or until the culture lyses.
9. Centrifuge for 1 min in a microfuge to pellet the cells and debris.
10. Pour the supernatant into a clean tube. Add several drops of chloroform and vortex.
11. Titer the phage on MS1883. Assign allele numbers to the mutants and save pickates of the mutant phage lysates at 4°. The DNA

sequence of the mutations can be determined by PCR using the anti-O_{mnt} and intra-Arc primers or by subcloning the EcoRI fragment that carries the O_{mnt} substitution and mnt::Kan onto a plasmid and sequencing the plasmid DNA.

Procedure 6. Construction of Challenge Phages Requiring Multiple Binding Sites

The challenge phage system can also be adapted to study mechanisms where DNA binding requires that two DNA-binding sites are bound, and efficient repression in the system results from binding at both sites and subsequent DNA looping through the interaction of proteins bound at both sites. In these constructs, the challenge phage system can be used to characterize the protein–protein interactions required for DNA looping (Nanassy and Hughes, 1998). Strain TH1901 is lysogenic for phage P22 Ap521. The P22 Ap521 has a transposon Tn1 inserted in the mnt gene that results in a phage that is too large to be packaged into a single P22 phage head. When plasmids that carry the mnt region of P22 are present in TH1901, induction of P22 Ap521 can result in recombination between the phage and the homologous DNA on the plasmid to yield phage recombinants that can be packaged into phage particles. Plasmid pMS284 (Youderian et al., 1983) is pBR322 based and contains a kanamycin-resistance cassette replacement of the mnt gene in a clone of the sieA - mnt - arc - ant region (immI) of P22 to yield sieA - neo - arc - ant where the neo sequence is flanked by PstI sites, and it also carries the $arcH1605^{Am}$ mutation. Induction of P22 Ap521 in the presence of pMS284 yields viable recombinants that are identical to P22-1000 described above. Another plasmid, pMS361, is identical to pMS284 except that the neo insert is in inverted orientation (Graña et al., 1988). In either construct, insertion of a DNA-binding sequence at the PstI site between the neo gene and sieA followed by recombination with P22 Ap521 after induction in strain TH1901 yields a parental challenge phage similar to P22-1000, except after recombination with pPY190 derivatives described above. The latter will result in challenge phages with two DNA-binding sites, one at Pant and a second about 1 kbp away between neo and sieA. These constructs are useful where DNA binding requires cooperative interactions between proteins bound at two nearby DNA sites and affect DNA looping between them. The same selections described above can be used to generate binding-site mutants in either site. These types of challenge phage were used to dissect the multiple steps in the Hin-mediated, site-specific recombination process in Salmonella (Nanassy and Hughes, 1998).

Procedure 7. Isolation of Challenge Phage with ant Expressed from Foreign Promoters

The challenge phage system can also be adapted to place the *ant* reporter under control of novel promoters by induction of the P22 lysogen in strain TH1901 in the presence of plasmid pMC16 derivatives. Plasmid pMC16 is a pBSIISK+ derivative with a 500-bp region of the *immI* region of P22 in which the *mnt* gene is replaced by a Km-resistance gene and by a multiple cloning site that replaces P*ant*. Induction of the P22 lysogen in TH1901 that also carries pMC16 derivatives can result in recombination between the sequence homology on the plasmid, and the resulting phages can be packaged into a single phage head and form a plaque on a P22-sensitive host. If promoter sequences are cloned into the multiple cloning site of pMC16 in the correct orientation, the resulting phage places the *ant* reporter under control of the novel promoter and Km^R as a selectable marker.

Acknowledgments

This work was supported by U.S. Public Health Service grants (GM34715, S.M. and GM28717, J.G.).

References

Ashraf, S., Kelly, M., Wang, Y., and Hoover, T. (1997). Genetic analysis of the *Rhizobium meliloti nifH* promoter, using the P22 challenge phage system. *J. Bacteriol.* **179,** 2356–2362.

Bass, S., Sugiono, P., Arvidson, D. N., Gunsalus, R. P., and Youderian, P. (1987). DNA specificity determinants of *Escherichia coli* tryptophan repressor binding. *Genes Dev.* **1,** 565–572.

Benson, N., Sugiono, P., Bass, S., Mendelman, L. V., and Youderian, P. (1986). General selection for DNA-binding activities. *Genetics* **118,** 21–29.

Benson, N., and Youderian, P. (1989). Phage λ Cro protein and cI repressor use two different patterns of specific protein-DNA interactions to achieve sequence specificity *in vivo.* *Genetics* **121,** 5–12.

Celander, D., Bennett, K., Fouts, D., Seitz, E., and True, H. (2000). RNA challenge phages as genetic tools for study of RNA-ligand interactions. *Methods Enzymol.* **318,** 332–350.

Chen, L., Goss, T., Bender, R., Swift, S., and Maloy, S. (1998). Genetic analysis, using P22 challenge phage, of the nitrogen activator protein DNA-binding site in the *Klebsiella aerogenes put* operon. *J. Bacteriol.* **180,** 571–577.

Chusacultanachai, S., Glenn, K., Rodriguez, A., Read, E., Gardner, J., Katzenellenbogen, B., and Shapiro, D. (1999). Analysis of estrogen response element binding by genetically selected steroid receptor DNA binding domain mutants exhibiting altered specificity and enhanced affinity. *J. Biol. Chem.* **274,** 23591–23598.

Fouts, D. E., and Celander, D. W. (1996). Improved method for selecting RNA-binding activities *in vivo.* *Nucleic Acids Res.* **24,** 1582–1584.

Gralla, J., and Collado-Vides, J. (1996). Organization and function of transcription regulatory elements. *In* "*Escherichia coli* and *Salmonella typhimurium*: Cellular and Molecular Biology" (F. C. Neidhardt, editor-in-chief). American Society for Microbiology, Washington, DC.

Graña, D., Gardella, T., and Susskind, M. M. (1988). The effects of mutations in the *ant* promoter of phage P22 depend on context. *Genetics* **120,** 319–327.

Hales, L., Gumport, R., and Gardner, J. (1994). Determining the DNA sequence elements required for binding integration host factor to two different target sites. *J. Bacteriol.* **176,** 2999–3006.

Han, Y., Gumport, R., and Gardner, J. (1994). Mapping the functional domains of bacteriophage lambda integrase protein. *J. Mol. Biol.* **235,** 908–925.

Hughes, K. T., Youderian, P., and Simon, M. I. (1988). Phase variation in *Salmonella*: Analysis of Hin recombination and *hix* recombination site interaction *in vivo. Genes Dev.* **2,** 937–948.

Lebreton, B., Prasad, P., Jayaram, M., and Youderian, P. (1988). Mutations that improve the binding of yeast FLP recombinase to its substrate. *Genetics* **118,** 393–400.

Lee, E., Gumport, R., and Gardner, J. (1990). Genetic analysis of bacteriophage λ integrase interactions with arm-type attachment site sequences. *J. Bacteriol.* **172,** 1529–1538.

Lee, E., Hales, L., Gumport, R., and Gardner, J. (1992). The isolation and characterization of mutants of the integration host factor (IHF) of *Escherichia coli* with altered, expanded DNA-binding specificities. *EMBO J.* **11,** 305–313.

Levine, M. (1957). Mutations in the temperate phage P22 and lysogeny in *Salmonella. Virology* **3,** 22–41.

MacWilliams, M., Gumport, R. I., and Gardner, J. F. (1997). Mutational analysis of protein binding sites involved in formation of the bacteriophage lambda *attL* complex. *J. Bacteriol.* **179,** 1059–1067.

MacWilliams, M. P., Celander, D. W., and Gardner, J. F. (1993). Direct genetic selection for a specific RNA-protein interaction. *Nucleic Acids Res.* **21,** 5754–5760.

Maloy, S., and Youderian, P. (1994). Challenge phage: Dissecting DNA–protein interactions *in vivo. Methods Genet.* **3,** 205–233.

Maloy, S. R., Stewart, V. J., and Taylor, R. K. (1996). "Genetic Analysis of Pathogenic Bacteria." Cold Spring Harbor Laboratory Press, Cold Spring Harbor, NY.

Nanassy, O. Z., and Hughes, K. T. (1998). *In vivo* identification of intermediate stages of the DNA inversion reaction catalyzed by the *Salmonella* Hin recombinase. *Genetics* **149,** 1649–1663.

Numrych, T. E., Gumport, R. I., and Gardner, J. F. (1991). A genetic analysis of Xis and Fis interactions with their binding sites in bacteriophage lambda. *J. Bacteriol.* **173,** 5954–5963.

Pfau, J., and Taylor, R. (1998). Mutations in *toxR* and *toxS* that separate transcriptional activation from DNA binding at the cholera toxin gene promoter. *J. Bacteriol.* **180,** 4724–4733.

Sanderson, K. E., and Hartman, P. E. (1978). Linkage map of *Salmonella typhimurium*, edition V. *Microbiol. Rev.* **42,** 471–519.

Slauch, J., Taylor, R., and Maloy, S. R. (1997). Survival in a cruel world: How *Vibrio cholerae* and *Salmonella* respond to an unwilling host. *Genes Dev.* **11,** 1761–1774.

Susskind, M. M., and Botstein, D. (1978). Molecular genetics of bacteriophage P22. *Microbiol. Rev.* **42,** 385–413.

Susskind, M. M., and Youderian, P. (1983). Bacteriophage P22 antirepressor and its control. *In* "Lambda II" (R. W. Hendrix J. W. Roberts F. W. Stahl and R. A. Weisberg, eds.). Cold Spring Harbor Laboratory Press, Cold Spring Harbor, NY.

Szegedi, S., and Gumport, R. (2000). DNA binding properties *in vivo* and target recognition domain sequence alignment analyses of wild-type and mutant *Rsr*I [N6-adenine] DNA methyltransferases. *Nucleic Acids Res.* **28**, 3972–3981.

Vogel, H. J., and Bonner, D. M. (1956). Acetylornithinase in *Escherichia coli*: Partial purification and some properties. *J. Biol. Chem.* **218**, 97–106.

Wang, S., True, H. L., Seitz, E. M., Bennett, K. A., Fouts, D. E., Gardner, J. F., and Celander, D. W. (1997). Direct genetic selection of two classes of R17/MS2 coat proteins with altered capsid assembly properties and expanded RNA-binding activities. *Nucleic Acids Res.* **25**, 1649–1657.

Youderian, P., Vershon, A., Bouvier, S., Sauer, R. T., and Susskind, M. M. (1983). Changing the DNA binding specificity of a repressor. *Cell* **35**, 777–783.

[19] Mud-P22

By JENNY A. CRAIN and STANLEY R. MALOY

Abstract

Mud-P22 derivatives are hybrids between phage Mu and P22 that can be inserted at essentially any desired site on the *Salmonella* chromosome (Benson and Goldman, 1992; Youderian *et al.*, 1988). Induction of Mud-P22 insertions yields phage particles that, as a population, carry chromosomal DNA from the region between 150 and 250 Kb on one side of the insertion. Thus, phage lysates from a representative set of Mud-P22 insertions into the *S. typhimurium* chromosome yield an ordered library of DNA that provides powerful tools for the genetic and physical analysis of the *Salmonella* genome. Although Mud-P22 has not yet been used in other species, this approach should be applicable in a variety of other bacteria as well.

Introduction

Phage Mu

Mu is a phage that replicates by transposition (Symonds *et al.*, 1987). Mu has a high transposition frequency and a low bias in selecting its transposition target site, so insertions can be easily recovered in any

METHODS IN ENZYMOLOGY, VOL. 421   0076-6879/07 $35.00
                    DOI: 10.1016/S0076-6879(06)21019-X

nonessential gene on the chromosome (Groisman, 1991). Transposition of Mu requires the Mu *A* and *B* gene products that act in *trans*, and sites at the left end (MuL) and right end (MuR) of the Mu genome that act in *cis* (Symonds *et al.*, 1987).

Deletion derivatives of Mu have been constructed (designated Mu*d*), some of which lack the Mu *A* and *B* genes that are required in *trans* for transposition, but retain the MuL and MuR sequences required in *cis* (Groisman, 1991). Such derivatives are unable to transpose unless the Mu *A* and *B* gene products are provided in *trans*. Delivery systems have been developed to allow a single transposition of the Mu*d* from the donor DNA by providing the Mu *A* and *B* gene products in *trans*. If the Mu *A* and *B* gene products are only provided transiently, no subsequent transposition can occur; thus, the resulting Mu*d* insertions are stable (Hughes and Roth, 1988).

Phage P22

Phage P22 is a temperate *Salmonella* phage that is very easy to grow and very stable during extended storage (Lawes and Maloy, 1993; Maloy *et al.*, 1996; Poteete, 1988; Susskind and Botstein, 1978). In many ways, P22 is quite similar to phage λ. Under conditions that favor lysogeny, P22 can integrate into the host genome at a specific attachment site via a site-specific recombination mechanism, catalyzed by the integrase protein (Int). Upon induction, excision of the prophage from the chromosome requires both the integrase and excisionase (Xis) proteins. In contrast to phage λ, packaging of P22 DNA into phage particles occurs by a headful mechanism.

Mu*d*-P22

Mu*d*-P22 is a hybrid between Mu and P22 phage (Fig. 1). One end of each Mu*d*-P22 contains 500 bp of the right end of Mu (MuR) and the other end contains 1000 bp from the left end of Mu (MuL). The central portion of Mu*d*-P22 contains the P22 genes required for replication and phage morphogenesis. However, Mu*d*-P22 derivatives are deleted for the P22 *attL*, *int*, *xis*, and *abc* genes at the left side of the prophage map, and *sieA*, gene *9*, *conABC*, and *attR* at the right end of the prophage map. The deletion of *attL*, *attR*, *int*, and *xis* prevents the integration or excision of Mu*d*-P22 via site-specific recombination. Thus, when integrated into the chromosome, the Mu*d*-P22 is "locked in." Deletion of the *abc* genes prevents a productive switch from bi-directional replication to rolling circle replication (Poteete, 1988). Deletion of *sieA* and *conABC* prevent

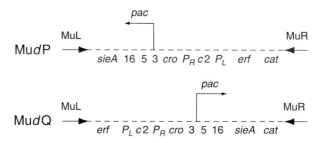

FIG. 1. Structure of the Mu*d*-P22 derivatives Mu*d*P and Mu*d*Q. The central portion of Mu*d*-P22 contains the P22 genome from the *erf* gene through part of the *sieA* gene. The opposite ends of Mu*d*-P22 contain approximately 500 bp of the right end of Mu (MuR) and 1000 bp from the left end of Mu (MuL). Since the orientation of the P22 segment within the Mu ends is opposite for Mu*d*P relative to Mu*d*Q, a Mu*d*P replacement of a Mu*d* insertion will package flanking DNA *either* clockwise *or* counterclockwise relative to genetic loci on the *S. typhimurium* genetic map, while the Mu*d*Q replacement of the same insertion will package genomic DNA in the opposite direction. Both Mu*d*P and Mu*d*Q are 32.4 Kb in size. A chloramphenicol resistance (Cam^R) gene is located adjacent to one end of Mu*d*-P22.

exclusion of superinfecting P22 phage. Thus, Mu*d*-P22 lysogens are still amenable to further genetic analysis using P22. Gene *9* encodes the tail fiber protein required for adsorption to recipient cells. Although phage that lack tail fibers are not infectious, tail protein can be extracted from cells containing a plasmid that expresses gene *9* constitutively and these "tails" will assemble onto P22 phage heads *in vitro*. The resulting "tailed" phage particles are fully infectious. Alternatively, a plasmid that expresses gene *9* can be introduced into the Mu*d*-P22 containing strain prior to induction so that tail fibers are provided in *trans* during the lytic development of the induced Mu*d*-P22 prophage.

Isolation of Chromosomal Mu*d*-P22 Insertions

Although it is possible to isolate chromosomal Mu*d*-P22 insertions by transposition if the Mu *A* and *B* gene products are provided in *trans* (Higgins and Hillyard, 1988; Youderian *et al.*, 1988), the frequency of transposition is much less than for the smaller Mu*d* derivatives. Therefore, chromosomal Mu*d*-P22 insertions are usually isolated by replacement of another chromosomal Mu*d* insertion via recombination *in vivo* (Fig. 2). Replacement of a chromosomal Mu*d* insertion with both a Mu*d*P and a Mu*d*Q will yield an isogenic pair of strains that package the chromosomal DNA in opposite directions (Fig. 2).

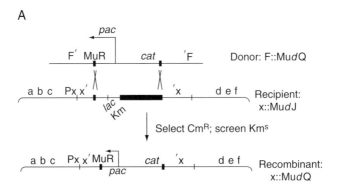

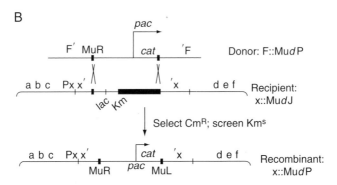

Fig. 2. Replacement of a chromosomal Mu*d* insertion with both Mu*d*P and a Mu*d*Q will yield an isogenic pair of strains that package the chromosomal DNA in opposite directions. Mu*d*-P22 insertions that package in a clockwise orientation are designated (A), and those that package in a counterclockwise orientation are designated (B), following the convention adopted for orientation of Tn*10* insertions (Chumley *et al.*, 1979). Since the packaging orientation of a Mu*d*-P22 is determined by the orientation of the Mu*d* element that was replaced, for a particular Mu*d* insertion, it is possible to find *either* a set of Mu*d*P(A) Mu*d*Q(B) *or* a set of Mu*d*P(B) Mu*d*Q(A) replacements, but not both sets. Any Mu*d* insertion that shares homology with Mu*d*-P22 at the flanking MuL and MuR sequences can be replaced via homologous recombination with Mu*d*P or Mu*d*Q.

Induction of Mu*d*-P22 Lysogens

DNA-damaging agents lead to inactivation of the P22 *c2* repressor, inducing the switch from lysogeny to lytic growth. For example, addition of 2 μg/ml mitomycin C to a mid-log phase culture of a P22 lysogen results in lysis and release of phage particles within 3 h. After excision, P22 normally circularizes, and then initiates bi-directional replication until about 20 copies of the phage genome are accumulated. In contrast, since "locked-in" prophages like Mud-P22 cannot excise, DNA replication begins from the P22 origin and extends beyond the ends of P22 into adjacent chromosomal DNA in both directions. This results in an "onion-skin" like amplification of the chromosome surrounding the "locked-in" prophage (Fig. 3).

While DNA replication is proceeding, expression of proteins required for phage morphogenesis also occurs. Once sufficient "Pac" nuclease (encoded by gene 3) has accumulated, the enzyme will cut a specific DNA sequence within gene 3, called the *pac* site. After cutting the DNA, the nuclease remains bound to the DNA and feeds it to the portal of the phagehead, where about 44 Kb of linear double-stranded DNA are stuffed into the capsid. Once the head is full, a second, nonspecific, cut is made by the nuclease to release the DNA, and assembly of the phage particle is completed. The end of DNA is then fed to another capsid and another headful is packaged. After about three to five consecutive, sequential "headfuls" of DNA are packaged, the enzyme dissociates from the DNA (Fig. 4). Only the initial cleavage by the nuclease for the first headful of DNA packaged is sequence dependent; all subsequent cuts occur at nonspecific sequences at the end of the DNA protruding from the capsid.

Since an induced Mud-P22 cannot excise from the chromosome, the first 44 Kb headful of DNA packaged will extend from the *pac* site through the right end of P22 and the flanking end of Mu (MuL for Mud*P* or MuR for Mud*Q*), and then into adjacent chromosomal DNA. The second and subsequent headfuls will exclusively contain sequential 44 Kb of chromosomal DNA. Although amplification is bi-directional, packaging proceeds in only one direction. Since Mud-P22 encodes the wild-type P22 gene 3 nuclease, the overwhelming majority of packaging events are initiated from the P22 *pac* site and not from pseudo-*pac* sites within the bacterial chromosome. Thus, the population of DNA molecules encapsidated upon induction of a specific Mud-P22 insertion is highly enriched for the 3 to 5 min of chromosomal DNA adjacent to one side of the Mud-P22 insertion.

Using Mud-P22 Insertions for Genetic and Physical Analysis of Chromosomal DNA

The phage particles obtained from a Mud-P22 insertion carry a specific 3 to 5 min region of adjacent chromosomal DNA. Thus, these phage particles provide a "specialized transducing lysate" for any specific region of the *S. typhimurium* chromosome. A set of Mud-P22 insertions spaced

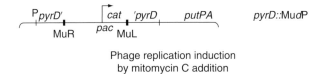

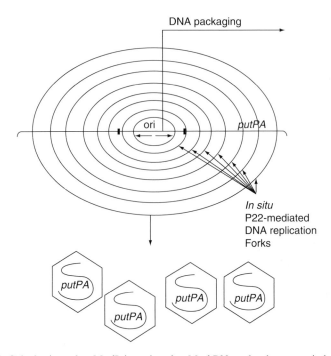

FIG. 3. Substitution of a MudJ insertion for Mud-P22 and subsequent induction and packaging of chromosomal DNA. An example showing the isolation of a *pyrD*::Mud-P22 at 24 min on the *S. typhimurium* chromosome and the subsequent induction of the Mud-P22–producing phage particles carrying the adjacent DNA (including the *putP* gene at about 25 min). A donor lysate of P22 is prepared on a strain containing F′ *zzf*::MudP or F′ *zzf*::MudQ, and the lysate used to transduce a Mu-containing recipient strain to Cam^R. The desired Cam^R recombinants lose the antibiotic resistance of the original Mud insertion (e.g., replacement of MudJ results in a Kan^S phenotype). Induction of the *pyrD*::Mud-P22 insertion with mitomycin C results in amplification of the adjacent chromosomal DNA, and packaging of three to five headfuls of DNA to one side of the *pac* site into P22 particles.

every 3 to 5 min around the *S. typhimurium* chromosome is available (Benson and Goldman, 1992; Maloy *et al.*, 1996). This ordered collection of Mud-P22 insertions allows the rapid genetic or physical mapping of genes on the *S. typhimurium* chromosome.

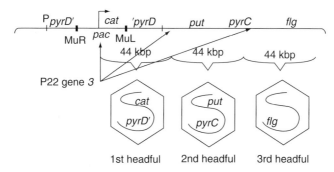

Fɪɢ. 4. Headful packaging of adjacent chromosomal DNA. Each phage head contains about 44 Kb of double-stranded linear DNA. The P22 gene 3 protein initially cuts the DNA at the *pac* site within the Mud-P22. One phage head is filled with DNA, and then the DNA is cut by the endonuclease. The next head is filled with DNA beginning with the resulting end. These reactions continue processively until three to five phage heads are filled with DNA; however, fewer phage particles are packaged with increasing distance from the *pac* site.

Rapid Genetic Mapping

The ordered set of "specialized" transducing lysates can be used for mapping the location of new mutations on the *S. typhimurium* chromosome by transduction. Recombinational repair of a chromosomal mutation by a particular Mud-P22 lysate will only occur if the corresponding wild-type allele is packaged into the transducing particles (i.e., located within 3 to 5 min in the proper orientation from the Mud-P22 insertion). The entire set of Mud-P22 lysates can be quickly tested by spot transduction of the mutant recipient with the Mud-P22 transducing lysates ordered as an array in a microtiter dish, and selecting for repair of the mutant phenotype. For this strategy to work, the recipient strain must be P22 sensitive and the mutant phenotype must be counter-selectable, so that only those recipient cells that are transduced to wild-type will form colonies on the selective plates. Some mutant phenotypes are inherently counter-selectable. For example, it is possible to select for repair of auxotropic mutations on minimal medium (Benson and Goldman, 1992), or temperature-conditional lethal mutations at the nonpermissive temperature (Gupta *et al.*, 1993). In addition, certain transposons carry genes with a counter-selectable phenotype. For example, the tetracycline-resistance gene present on transposon Tn*10* makes enteric bacteria sensitive to certain lipophilic chelating agents (Bochner *et al.*, 1980; Maloy and Nunn, 1981), and the *Bacillus subtilis sacB* (secretory levansucrase) gene present on the MudSacI transposon makes gram-negative enteric bacteria sensitive to 5% sucrose (Lawes and Maloy, 1995). It is possible to directly select for tetracycline sensitivity due to loss of a Tn*10* insertion, or for sucrose resistance due to loss of a Mu*d*SacI insertion. Consequently, if a mutation does not have a counter-selectable phenotype, it is straightforward to isolate a closely linked Tn*10* or Mu*d*SacI insertion whose location can be mapped by transduction versus the collection of Mu*d*-P22 lysates (Maloy *et al.*, 1996).

An alternative strategy is to generate mutations with Mu*d* insertions and then to replace each insertion with Mu*d*-P22, or to isolate Mu*d*-P22s linked to the gene of interest. Induction of the resulting Mu*d*-P22 yields a "specialized" transducing lysate that can be used to transduce an ordered set of auxotrophs, selecting for repair of the auxotropic mutation on minimal media (Higgins and Hillyard, 1988). Since most of the DNA in the transducing particles is from the 3-to-5–min region adjacent to the Mu*d*-P22 insertion, only those auxotrophic mutations that are near the Mu*d*-P22 insertion will be repaired at high frequency. Thus, the approximate location of the Mu*d*-P22 insertion (and the corresponding Mu*d* insertion) can be inferred from the transduction frequency. This approach is conceptually similar to the use of directed Hfr formation to determine the location of Tn*10* insertions (Chumley *et al.*, 1979).

Cloning, Restriction Mapping, and DNA Sequencing

Mu*d*-P22 lysates can also be used as a highly enriched source of DNA for cloning, restriction mapping, or sequencing genes from the *S. typhimurium* chromosome. If the approximate map location of a mutation is known, then a closely linked Mu*d*-P22 with the appropriate packaging direction can be induced and DNA extracted from the phage used to clone the gene of interest (Higgins and Hillyard, 1988).

The DNA can also be directly used for restriction mapping (Higgins and Hillyard, 1988; Youderian *et al.*, 1988) or DNA sequencing without an intermediate cloning step. For example, Hughes *et al.* (1993) isolated 6 MuJ insertions located at different positions in the *nadC* gene, replaced each of the insertions with a Mu*d*-P22, and then isolated DNA from each of the Mu*d*-P22 lysates. The DNA was directly sequenced using primers specific to one of the ends of Mu (MuL for Mu*d*P and MuR for Mu*d*Q) to read from the flanking Mu end into the adjacent *nadC* DNA.

DNA Hybridization

DNA prepared from the representative set of Mu*d*-P22 lysates also provides an ordered array of chromosomal DNA fragments for physical mapping by DNA hybridization. A single hybridization filter, with an ordered pattern of spots of DNA from a representative set of Mu*d*-P22 insertions, can be probed with smaller chromosomal fragments, cloned genes, or cloned cDNA corresponding to differentially expressed genes (Libby *et al.*, 1994; Wong and McClelland, 1992, 1994; Wong *et al.*, 1994). Applications of this approach are analogous to those used for the Kohara library from *Escherichia coli* (Kohara *et al.*, 1987).

DNA purified from a particular Mu*d*-P22 lysate can be used to probe a Southern blot. For example, Liu and Sanderson (1992, chapter 24) used DNA purified from Mu*d*-P22 lysates to probe digests of *S. typhimurium* total chromosomal DNA separated by pulsed-field gel electrophoresis. This approach facilitated the compilation of a correlated physical–genetic map of *S. typhimurium* (see map 33, section 3, Sanderson *et al.*, 1995).

Use of Mud-P22 in Other Species

Mu*d*-P22 can be used in any strain of *Salmonella* that is sensitive to P22. Although Mu*d*-P22 has only been used in *Salmonella* thus far, with some modifications this system should be applicable in a variety of other bacteria as well. For example, a cosmid that carries the *S. typhimurium* lipopolysaccharide genes renders some gram-negative bacteria sensitive to P22 (Neal *et al.*, 1993). Even if Mu does not transpose in the bacterium, Mu*d*-P22 insertions could be obtained by mating an F′ carrying Mu*d*-P22 into the bacterium and selecting for homologous recombination between a transposon present both on the F′ and the chromosome (such as Tn*10*) or by recombination between any gene on the F′ and the corresponding chromosomal gene (Zahrt and Maloy, submitted; Zahrt *et al.*, 1994). Alternatively, it should be possible to develop analogous approaches using phage specific for any particular bacterium.

Summary

Mu*d*-P22 derivatives are hybrids between phage Mu and phage P22 that as a population can package 150 to 250 Kb of DNA from any specific region of the *Salmonella* chromosome. An ordered set of Mu*d*-P22 insertions spaced every 3 to 5 min around the chromosome allows the rapid genetic mapping, physical mapping, cloning, and sequencing of genes from *S. typhimurium*. Thus, Mu*d*-P22 derivatives provide powerful tools for the genetic and physical analysis of the *Salmonella* genome, and maybe a useful tool for the genetic and physical analysis of other bacterial species as well.

References

Benson, N. R., and Goldman, B. S. (1992). Rapid mapping in *Salmonella typhimurium* with Mu*d*-P22 prophages. *J. Bacteriol.* **174,** 1673–1681.

Bochner, B., Huang, H. C., Schrevin, G. L., and Ames, B. N. (1980). A positive selection for loss of tetracycline resistance. *J. Bacteriol.* **143,** 926–933.

Chumley, F. G., Menzel, R., and Roth, J. R. (1979). Hfr formation directed by Tn*10. Genetics* **91,** 639–655.

Groisman, E. A. (1991). *In vivo* genetic engineering with bacteriophage Mu. *Methods Enzymol.* **204,** 180–212.

Gupta, S. D., Gan, K., Schmid, M. B., and Wu, H. C. (1993). Characterization of a temperature-sensitive mutant of *Salmonella typhimurium* defective in apolipoprotein N-acetyltransferase. *J. Biol. Chem.* **268,** 16551–16556.

Higgins, N. P., and Hillyard, D. (1988). Primary structure and mapping of the *hupA* gene of *Salmonella typhimurium. J. Bacteriol.* **170,** 5751–5758.

Hughes, K. T., Dessen, A., Gray, J. P., and Grubmeyer, C. (1993). The *Salmonella typhimurium nadC* gene: Sequence determination by use of Mu*d*-P22 and purification of quinolate phosphoribosyltransferase. *J. Bacteriol.* **175,** 479–486.

Hughes, K. T., and Roth, J. R. (1988). Transitory *cis* complementation: A method for providing transposition functions to defective transposons. *Genetics* **119,** 9–12.

Kohara, Y., Akiyama, K., and Isono, K. (1987). The physical map of the whole *E.coli* chromosome: Application of a new strategy for rapid analysis and sorting of a large genomic library. *Cell* **50,** 495–508.

Lawes, M. C., and Maloy, S. R. (1993). Genetics of DNA injection by phages λ and P22. *Curr. Top. Mol. Genet.* **1,** 133–146.

Lawes, M. C., and Maloy, S. R. (1995). Mu*d*SacI, a transposon with strong selectable and counterselectable markers: Use for rapid mapping of chromosomal mutations in *Salmonella typhimurium. J. Bacteriol.* **177,** 1383–1387.

Libby, S. J., Goebel, W., Ludwig, A., Buchmeier, N., Bowe, F., Fang, F. C., Guiney, D. G., Songer, J. G., and Heffron, F. (1994). A cytolysin encoded by *Salmonella* is required for survival within macrophages. *Proc. Natl. Acad. Sci. USA* **91,** 489–493.

Liu, S. L., and Sanderson, K. E. (1992). A physical map of the *Salmonella typhimurium* LT2 genome made by using *XbaI* analysis. *J. Bacteriol.* **174,** 1662–1672.

Maloy, S. R., and Nunn, W. D. (1981). Selection for loss of tetracycline resistance by *Escherichia coli. J. Bacteriol.* **145,** 1110–1112.

Maloy, S. R., Stewart, V. J., and Taylor, R. K. (1996). "Genetic Analysis of Pathogenic Bacteria: A Laboratory Manual." Cold Spring Harbor Laboratory Press, Cold Spring Harbor, NY.

Neal, B. L., Brown, P. K., and Reeves, P. R. (1993). Use of *Salmonella* phage P22 for transduction in *Escherichia coli. J. Bacteriol.* **175,** 7115–7118.

Poteete, A. R. (1988). Bacteriophage P22. *In* "The Bacteriophages" (R. Calendar, ed.), Vol. 2. Plenum Press, New York.

Sanderson, K. E., Hessel, A., and Rudd, K. E. (1995). Genetic map of *Salmonella typhimurium*, edition VIII. *Microbiol. Rev* **59,** 241–303.

Susskind, M. M., and Botstein, D. (1978). Molecular genetics of bacteriophage P22. *Microbiol. Rev.* **42,** 385–413.

Symonds, N., Toussaint, A., van de Putte, P., and Howe, M. M. (eds.) (1987)."Phage Mu." Cold Spring Harbor Laboratory Press, Cold Spring Harbor, NY.

Wong, K. K., and McClelland, M. (1992). A *BlnI* restriction map of the *Salmonella typhimurium* LT2 genome. *J. Bacteriol.* **174,** 1656–1661.

Wong, K. K., and McClelland, M. (1994). Stress-inducible gene of *Salmonella typhimurium* identified by arbitrarily primed PCR of RNA. *Proc. Natl. Acad. Sci. USA* **91,** 639–643.

Wong, K. K., Wong, R. M., Rudd, K. E., and McClelland, M. (1994). High-resolution restriction map for a 240-kilobase region spanning 91 to 96 minutes on the *Salmonella typhimurium* LT2 chromosome. *J. Bacteriol.* **176,** 5729–5734.

Youderian, P., Sugiono, P., Brewer, K. L., Higgins, N. P., and Elliott, T. (1988). Packaging specific segments of the *Salmonella* chromosome with locked-in Mu*d*-P22 prophages. *Genetics* **118,** 581–592.

Zahrt, T. C., Mora, G. C., and Maloy, S. (1994). Inactivation of mismatch repair overcomes the barrier to transduction between *Salmonella typhimurium* and *Salmonella typhi*. *J. Bacteriol.* **176,** 1527–1529.

[20]   Phage Metagenomics

By VERONICA CASAS and FOREST ROHWER

Abstract

The vast majority of novel DNA sequences deposited in the databases now comes from environmental phage DNA sequences. Methods are presented for the cloning and sequencing of phage DNA that might otherwise be lethal to bacterial host vectors or contain modified DNA bases that prevent standard cloning of such sequences. In addition, methods are presented for the isolation of viral particles directly from soil and sediment environmental samples or from large volumes of environmental water samples. The viral particles are then purified by cesium-chloride density centrifugation followed by DNA extraction. This purified viral metagenomic DNA is then used for cloning and sequencing.

Introduction

Bacterial viruses (bacteriophage or simply phage) are the most abundant biological entities on the planet. There are approximately 10^6 phage per milliliter in the world's oceans and lakes and 10^9 phage per gram of sediment and topsoil (Bergh *et al.*, 1989; Danovaro and Serresi, 2000; Hewson *et al.*, 2001; Maranger and Bird, 1996; Ogunseitan *et al.*, 1990). Phage are the major predators of bacteria and are believed to influence the types and population density of bacteria in an environment. By killing bacteria, phage modulate global biogeochemical cycles, an example of which is the marine microbial

METHODS IN ENZYMOLOGY, VOL. 421   0076-6879/07 $35.00
                    DOI: 10.1016/S0076-6879(06)21020-6

food web (Azam, 1988; Azam and Ammerman, 1984; Azam et al., 1983). Phage have also been implicated in the maintenance of microbial diversity by selective killing (Bratbak et al., 1992; Fuhrman, 1999; Thingstad et al., 1993; Wommack and Colwell, 2000).

In addition to controlling bacterial populations by lysis of infected bacteria, phage can also alter the physiology of infected bacteria through horizontal gene transfer. Many temperate phage express gene products that alter the phenotype of the bacterial host through lysogenic conversion. One of the most common examples of lysogenic conversion is immunity to superinfection by other phage. Lysogenic conversion can also result in expanded metabolic capabilities including resistance to antibiotics and reactive oxygen compounds (Mlynarczyk et al., 1997; Ochman et al., 2000). For phage that carry exotoxin genes, lysogenic conversion can change avirulent bacteria into human pathogens (Banks et al., 2002; Canchaya et al., 2003, 2004).

Studies have also suggested that phage can readily move between different types of ecosystems and have the ability to infect bacteria from these ecosystems. In a study surveying the distribution of T7-like podophage, DNA polymerase genes these sequences were found to occur in marine, freshwater, sediment, terrestrial, extreme, and metazoan-associated ecosystems (Breitbart and Rohwer, 2004). This suggested that the phage have moved in relatively recent evolutionary time (Breitbart and Rohwer, 2004). Moreover, it has also been shown that phage from one type of ecosystem (e.g., soil, sediment, or freshwater) can grow on bacterial hosts isolated from a distinctly different ecosystem (e.g., marine environments [Sano et al., 2004]). Together these results implied that phage from various ecosystem types are capable of infecting more than one type of bacterial host and, as a result, are capable of moving DNA between these ecosystem types.

Considering the impact that phage have on biogeochemical cycling, bacterial population densities and community structure, horizontal gene transfer, and bacterial virulence, it is surprising that more is not known about their diversity and biogeography. The standard methods used to study their bacterial counterparts cannot be directly applied to the study of phage diversity and biogeography. Phage DNA is lethal to bacterial cells and it contains modified nucleotide bases that are a barrier to standard cloning techniques (Wang et al., 2000; Warren, 1980; Xu et al., 2002). Development of novel cloning and sequencing techniques have overcome these obstacles and viral metagenomics is beginning to provide a better understanding of the ecology of viruses (Breitbart et al., 2002; Margulies et al., 2005; Rohwer et al., 2001).

The term "viral metagenomics" can be defined as the culture-independent functional and sequence-based analysis of an assemblage of phage genomes in an environmental sample (Handelsman et al., 1998; Riesenfeld et al., 2004).

To date, there have been five dsDNA and two RNA viral metagenomic libraries published. The dsDNA viral metagenomic libraries included viruses from two near-shore marine water samples (Breitbart *et al.*, 2002), one marine sediment sample, one human fecal sample (Breitbart *et al.*, 2003), and one equine fecal sample (Cann *et al.*, 2005). The two RNA viral metagenomic libraries were derived from viruses isolated from coastal waters off of Canada (Culley *et al.*, 2006) and from human feces (Zhang *et al.*, 2006). What was overwhelming in these studies was that the vast majority of viral sequences showed no significant similarity (E-value >0.001) to sequences deposited in the GenBank nonredundant database (Edwards and Rohwer, 2005). The knowledge of the community structure and composition of uncultured microbes has grown dramatically via the use of metagenomics, and viral metagenomics is likely to provide similar insights into the number and types of phage in the environment.

Procedures

The methods described here focus on the isolation, purification, and extraction of DNA from double-stranded DNA viruses. Appropriate modifications are needed for cloning and sequencing single-stranded DNA and RNA viruses (see Culley *et al.*, 2006; Zhang *et al.*, 2006). Random-primed reverse transcriptase and strand displacement DNA polymerases may be viable options for these types of viruses (Edwards and Rohwer, 2005).

Protocol 1. Isolation of Viral Particles from Soil and Sediment Environmental Samples

1. For soil samples:

   a. In a sterile appropriately sized container, add equal amounts of $1\times$ SM buffer and soil sample volumes (i.e., 50 ml buffer to 50 g soil).

   b. Shake vigorously for a few minutes until soil is well suspended and to ensure that viral particles are released from the soil sample.

   c. Continue to step 3.

2. For sediment samples (freshwater or salt water):

   a. In a sterile appropriately sized container, add equal volumes of sediment sample and 0.2 μm filtered, autoclaved water from the sample location.

   b. Shake vigorously for a few minutes until sediment is well suspended and to ensure that viral particles are released from the sediment sample.

   c. Continue to step 3.

3. Allow soil/sediment to settle to the bottom of the container at 4°. This may take a few hours, so it may be best to store at 4° overnight.

4. Using a sterile pipette, transfer the supernatant to an appropriately sized sterile container for centrifugation. Conical or centrifuge tubes work best for this step.

5. Spin for 15 min at the highest speed allowed for the container and centrifuge you are using (10,000g if allowable). This will pellet any soil/sediment debris remaining in your supernatant.

6. Pour supernatant into sterile 60-ml syringe with 0.2 μm Sterivex (Millipore, Billerica, MA) filter attached to the tip.

7. Filter supernatant into an appropriately sized sterile conical or centrifuge tube.

8. To the filtrate add ~10 units or Kunitz units of DNase I per milliliter of filtrate. This amount of DNase I can be increased, or the treatment repeated, if all the free DNA is not removed from the filtrate.

9. Incubate at room temperature for 1 h.

10. Precipitate viral particles by adding 10% (w/v) solid polyethylene glycol (PEG) 8000. Make sure that the PEG is well dissolved. For best results, precipitate overnight at 4°.

11. Centrifuge sample with PEG for 15 to 30 min at 11,000g.

12. Decant the supernatant. The viral particles are in the pellet.

13. Invert the conical or centrifuge tube containing the pellet in a tilted position for 5 min to remove any residual liquid.

14. Add the desired/necessary volume of TE buffer (pH 7.6) to the viral pellet. Let stand for a few minutes at room temperature, and then resuspend the pellet with a wide-bore pipette. Do this gently, as the viruses can be sheared if pipetted too violently.

15. Transfer suspension to new appropriately sized container.

16. DNA can now be extracted from the pelleted viral particles using protocol 4. If contamination with exogenous DNA is a problem, the viral particles may first be further purified using pro- tocol 3 and then the DNA extracted. (This protocol was adapted from Maniatis et al., 1982; Sambrook et al., 1989; and Sambrook and Russell, 2001.)

Protocol 2. Isolation of Viral Particles from Large Water Environmental Samples Using Tangential-Flow Filtration

1. Using a vacuum pump, first filter the sample through a 0.45 μm GF/F filter (Whatman Inc.; Florham Park, NJ) to remove protists and large bacteria from the sample. Repeat this step to ensure that all

protists are removed (Wilcox and Fuhrman, 1994). This is the 0.45 μm filtrate.

2. Set up the tangential flow-filter (TFF) system. The main components are the TFFs (Amersham Biosciences, Piscataway, NJ), peristaltic pump, tubing to connect the reservoirs to the filter and pump, and reservoirs to contain the sample as well as to capture the filtrate. The reservoir can be any container ranging from a beaker to a large trashcan.

3. Also needed are pressure gauges to monitor the pressure within the TFF system. It is important that the correct amount of pressure is maintained within the system such that the sample is forced through the TFF pores, while at the same time not exceeding 10 psi so that the viral particles do not burst.

4. Run the 0.45-μm filtrate through a 0.2-μm TFF. The 0.2-μm TFF removes bacteria, but allows viruses to pass through.

5. Circulate the entire 0.45-μm filtrate through the filter until it is concentrated down to a volume of $\sim$1 l. This is the 0.2-μm filtrate.

6. Run the 0.2-μm filtrate through a 100-kDa TFF. Viral particles will either be trapped on the filter or recycled back into the sample reservoir. The filter is run until there is very little volume left in the sample reservoir. At this point, the pressure on the retentate tube is released, allowing the sample to wash over the filter, freeing any attached viruses. The sample is then concentrated further, with the final step of removing the input tube that runs air through the filter and forces all remaining sample out of the filter and tubing. The viral concentrate is in the sample reservoir.

7. This method will concentrate volumes >10 l down to $\sim$1 l, and volumes less than 10 l down to $\sim$100 ml.

8. Viral particles can now be further purified using protocol 3. (This protocol was based on Breitbart *et al.*, 2002, 2004a,b; Sano *et al.*, 2004; Wommack *et al.*, 1995).

Protocol 3. Purification of Viral Particles by Cesium Chloride (CsCl) Density Centrifugation

1. To the viral concentrate, add 0.2 g CsCl per milliliter of viral concentrate.

2. Make three CsCl solutions of 1.35-g/ml, 1.5-g/ml, and 1.7-g/ml densities. Make the CsCl gradient from the same solution in which the viral concentrate is diluted, and make sure that it has been filtered with a 0.02-μm filter to remove any possible contaminating viral particles.

3. Set up the CsCl gradient in clear plastic centrifuge tubes that fit the rotor of the high-speed centrifuge being utilized (e.g., Beckman Ultra-Clear centrifuge tubes).

4. Layer the three CsCl solutions from greatest density to least density by slowly trickling the solution down the side of the centrifuge tube, being careful not to mix the gradients in the tube. The CsCl step gradient should take up ~60% of the centrifuge tube. The number of step gradients required to purify the entire viral concentrate is the final aqueous volume from step 1 divided by 40% of the volume of the centrifuge tube.

5. As you add each solution, mark the outside of the tube to denote the location of each fraction.

6. Carefully layer the appropriate volume of viral concentrate on top of the gradient (~40% of the capacity of the centrifuge tube).

7. Load an even number of step gradients into the centrifuge, making sure that opposite tubes are carefully balanced.

8. Centrifuge the gradients at 87,000g for 2 h at 4°.

9. Remove the gradients from the centrifuge and wipe the outside of each tube with ethanol to remove grease or oils. Apply a piece of clear tape to the outside of the tube level with the 1.35-g/ml and 1.5-g/ml densities.

10. Use a 21-gauge needle to pierce the tube, through the tape, just below the 1.5-g/ml density. Be careful to keep fingers away from the other side of the tube, just in case the needle punctures all the way through the tube.

11. Collect the 1.5-g/ml density and the interface between the 1.35-g/ml and 1.5-g/ml densities.

12. Extract DNA from the viral particles using protocol 4. (This protocol was adapted from Maniatis *et al.*, 1982; Sambrook and Russell, 2001; Sambrook *et al.*, 1989.)

Protocol 4. Extraction of DNA from Viral Particles Using Formamide and CTAB/NaCl

Formamide Preparation

1. To the viral concentrate, add the following:
   a. 0.1 volume of 2 M Tris-Cl (pH 8.5)/0.2 M EDTA
   b. 0.05 volume of 0.5 M EDTA
   c. 1 volume of deionized formamide
   d. 10 μl glycogen (10 mg/ml)
2. Incubate at room temperature for 30 min.
3. Add 2 volumes of room temperature 100% ethanol.
4. Incubate overnight at –20°.

5. Centrifuge at high speed (e.g., 10,000g) for 20 min at 4°.
6. Decant supernatant.
7. Add 500 μl of cold 70% ethanol to wash the pellet.
8. Centrifuge at high speed for 10 min.
9. Decant supernatant and repeat wash step.
10. Resuspend into 567 μl TE buffer (pH 8.0). Be careful not to vortex or pipette too vigorously.
11. Continue with "CTAB/NaCl preparation" protocol below.

CTAB/NaCl Preparation

1. To 567 μl of the resuspended viral pellet, add 30 μl SDS (0.5% final concentration) and 3 μl proteinase K (100 μg/ml final concentration). Mix.
2. Incubate for 1 h at 37°.
3. Add 100 μl of 5M NaCl to the resuspended viral pellet and mix thoroughly by inversion. Ensure that the final NaCl concentration is >0.5 M so that the nucleic acid does not precipitate.
4. Add 80 μl CTAB/NaCl solution to the resuspended viral pellet and mix thoroughly by inversion.
5. Incubate for 10 min at 65°.
6. Add an equal volume of 24:1 chloroform/isoamyl alcohol and mix thoroughly by inversion.
7. Centrifuge for 5 min at high speed.
8. Transfer supernatant to a new tube, being careful not to transfer the debris in the interface. DNA is in the supernatant.
9. Add equal volume of 25:24:1 phenol/chloroform/isoamyl alcohol to the supernatant and mix thoroughly by inversion.
10. Centrifuge for 5 min at high speed.
11. Transfer supernatant to a new tube.
12. Add 0.7 volume isopropanol to the supernatant fraction, and mix gently by rocking the tube parallel to the ground until a white DNA precipitate forms.
13. Centrifuge at high speed for 15 min at 4°.
14. Decant supernatant being careful not to discard the pellet.
15. Add 500 μl 70%-ethanol to the pellet.
16. Centrifuge for 5 min at high speed. Decant ethanol.
17. Repeat steps 15 and 16.
18. Remove all residual ethanol with a pipette and let DNA pellet air dry, or dry pellet using a lyophilizer.
19. Resuspend pellet in 50 μl sterile, DNAse-, RNAse-free water.
20. This purified viral metagenomic DNA can be used for cloning and sequencing. Many phage modify their DNA in ways that complicate

traditional cloning approaches. The phage DNA can be cloned and sequenced as described by Rohwer *et al.* (2001). Alternatively, a more economical solution is to take advantage of companies offering technologies that circumvent problems with cloning phage DNA. Two solutions are the PicoTiterPlate technology from 454 Life Sciences (Branford, CT), the Linker-Amplified-Shotgun-Libraries (LASLs) from Lucigen Corporation (Middleton, WI), or both.

Reagents

CTAB/NaCl (100 ml)

1. Add 4.1 g NaCl to 80 ml water.
2. Slowly add 10 g cetyltrimethyl ammonium bromide (CTAB) while stirring (heat to 65° if necessary).
3. Bring volume up to 100 ml.

5× Storage Media (SM) Buffer (1 L)

1. Weigh out 29 g NaCl and 10 g $MgSO_4$ * $7H_2O$.
2. Measure out 250 ml 1 *M* Tris-Cl at pH 7.5.
3. Add water to a final volume of 1 l. Make appropriate dilution for working stock of 1X. (This protocol was adapted from Ausubel *et al.*, 1995, 2002; Maniatis *et al.*, 1982; Sambrook and Russell, 2001; Sambrook *et al.*, 1989.)

Acknowledgments

This work was done in collaboration with Dr. Forest Rohwer. This work is supported by NSF grant (DEB03-16518) and the Gordon and Betty Moore Foundation to Forest Rohwer. Veronica Casas was funded by a fellowship from NIH/NIGMS Minority Biomedical Research Support program (5R25-GM8907).

References

Ausubel, F., Brent, R., Kingston, R. E., Moore, D. D., Seidman, J. G., Smith, J. A., and Struhl, K. (eds.) (1995). "Short Protocols in Molecular Biology," 3rd ed. John Wiley and Sons, New York.

Ausubel, F., Brent, R., Kingston, R. E., Moore, D. D., Seidman, J. G., Smith, J. A., and Struhl, K. (eds.) (2002). "Short Protocols in Molecular Biology," 5th ed. John Wiley and Sons, New York.

Azam, F. (1988). "The Microbial Food Web in the Sea." Marine Science of the Arabian Sea. Proceedings of an International Conference.

Azam, F., and Ammerman, J. W. (1984). Cycling of organic matter by bacterioplankton in pelagic marine ecosystems: Microenvironmental considerations. *In* "Flows of Energy and Materials in Marine Ecosystems" (M. J. R. Fasham, ed.). Plenum, New York.

Azam, F., Fenchel, T., Field, J. G., Gray, J. S., Meyer-Reil, L. A., and Thingstad, F. (1983). The ecological role of water-column microbes in the sea. *Mar. Ecol. Prog. Ser.* **10,** 257–263.

Banks, D. J., Beres, S. B., and Musser, J. M. (2002). The fundamental contribution of phages to GAS evolution, genome diversification and strain emergence. *Trends Microbiol.* **10,** 515–521.

Bergh, O., Borsheim, K. Y., Bratback, G., and Heldal, M. (1989). High abundance of viruses found in aquatic environments. *Nature* **340,** 467–468.

Bratbak, G., Heldal, M., Thingstad, T. F., Riemann, B., and Haslund, O. H. (1992). Incorporation of viruses into the budget of microbial C-transfer. A first approach. *Mar. Ecol. Prog. Ser.* **83,** 273–280.

Breitbart, M., Felts, B., Kelley, S., Mahaffy, J. M., Nulton, J., Salamon, P., and Rohwer, F. (2004a). Diversity and population structure of a nearshore marine sediment viral community. *Proc. R. Soc. Lond. B Biol. Sci.* **271,** 565–574.

Breitbart, M., Hewson, I., Felts, B., Mahaffy, J. M., Nulton, J., Salamon, P., and Rohwer, F. (2003). Metagenomic analyses of an uncultured viral community from human feces. *J. Bacteriol.* **85,** 6220–6223.

Breitbart, M., and Rohwer, F. (2004). Global distribution of nearly identical phage-encoded DNA sequences. *FEMS Microbiol. Lett.* **236,** 245–252.

Breitbart, M., Salamon, P., Andresen, P., Mahaffey, J. M., Segall, A. M., Mead, D., Azam, F., and Rohwer, F. (2002). Genomic analysis of uncultured marine viral communities. *Proc. Natl. Acad. Sci. USA* **99,** 14250–14255.

Breitbart, M., Wegley, W., Leeds, S., Schoenfeld, T., and Rohwer, F. (2004b). Phage community dynamics in hot springs. *Appl. Eviron. Microbiol.* **70,** 1633–1640.

Canchaya, C., Fournous, G., and Brussow, H. (2004). The impact of prophages on bacterial chromosomes. *Mol. Microbiol* **53,** 9–18.

Canchaya, C., Fournous, G., Chibani-Chennoufi, S., Dillman, M. L., and Brussow, H. (2003). Phage as agents of lateral gene transfer. *Curr. Opin. Microbiol.* **6,** 417–424.

Cann, A. J., Fandrich, S. E., and Heaphy, S. (2005). Analysis of the virus population present in equine faeces indicates the presence of hundreds of uncharacterized virus genomes. *Virus Genes* **30,** 151–156.

Culley, A. I., Lang, A. S., and Suttle, C. A. (2006). Metagenomic analysis of coastal RNA virus communities. *Science* **312,** 1795–1798.

Danovaro, R., and Serresi, M. (2000). Viral density and virus-to-bacterium ratio in deep-sea sediments of the Eastern Mediterranean. *Appl. Environ. Microbiol.* **66,** 1857–1861.

Edwards, R. A., and Rohwer, F. (2005). Viral metagenomics. *Nat. Rev. Microbiol.* **3,** 504–510.

Fuhrman, J. A. (1999). Marine viruses: Biogeochemical and ecological effects. *Nature* **399,** 541–548.

Handelsman, J., Rondon, J., Brady, M., Clardy, J., and Goodman, R. M. (1998). Molecular biological access to the chemistry of unknown soil microbes: A new frontier for natural products. *Chem. Biol.* **5,** R245–R249.

Hewson, I., O'Neil, I., Fuhrman, J. A., and Dennison, W. C. (2001). Virus-like particle distribution and abundance in sediments and overlying waters along eutrophication gradients in two subtropical estuaries. *Limnol. Oceanogr.* **46,** 1734–1746.

Maniatis, T., Fritsch, E. F., and Sambrook, J. (1982). "Molecular Cloning: A Laboratory Manual." Cold Spring Harbor Laboratory Press, Cold Spring Harbor, NY.

Maranger, R., and Bird, D. F. (1996). High concentrations of viruses in the sediments of Lac Gilbert, Quebec. *Microb. Ecol.* **31,** 141–151.

Margulies, M., Egholm, M., Altman, W. E., Attiya, S., Bader, J. S., Bemben, L. A., Berka, J., Braverman, M. S., Chen, Y.-J., Chen, Z., Dewell, S. B., Du, L., *et al.* (2005). Genome sequencing in microfabricated high-density picolitre reactors. *Nature* **437,** 326–327.

Mlynarczyk, G., Mlynarcyk, A., Zabicka, D., and Jeljaszewicz, J. (1997). Lysogenic conversion as a factor influencing the vancomycin tolerance phenomenon in *Staphylococcus aureus*. *J. Antimicrob. Chemother.* **40,** 136–137.

Ochman, H., Lawrence, J. G., and Groisman, E. (2000). Lateral gene transfer and the nature of bacterial innovation. *Nature (Lond.)* **405,** 299–304.

Ogunseitan, O. A., Sayler, G. S., and Miller, R. V. (1990). Dynamic interactions of *Pseudomonas aeruginosa* and bacteriophages in lake water. *Microb. Ecol.* **19,** 171–185.

Riesenfeld, C. S., Schloss, P. D., and Handelsman, J. (2004). Metagenomics: Genomic analysis of microbial communities. *Annu. Rev. Genet.* **38,** 525–552.

Rohwer, F., Seguritan, V., Choi, D. H., Segall, A. M., and Azam, F. (2001). Production of shotgun libraries using random amplification. *BioTechniques* **31,** 108–118.

Sambrook, J., and Russell, D. W. (2001). "Molecular Cloning: A Laboratory Manual." Cold Spring Harbor Laboratory Press, Cold Spring Harbor, NY.

Sambrook, J., Fritsch, E., and Maniatis, T. (1989). "Molecular Cloning: A Laboratory Manual." Cold Spring Harbor Laboratory Press, Cold Spring Harbor, NY.

Sano, E., Carlson, S., Wegley, L., and Rohwer, F. (2004). Movement of viruses between biomes. *Appl. Environ. Microbiol.* **70,** 5842–5846.

Thingstad, T. F., Heldal, M., Bratback, G., and Dundas, I. (1993). Are viruses important partners in pelagic food webs? *Trends Ecol. Evol.* **8,** 209–213.

Wang, I., Smith, D., and Young, R. (2000). Holins: The protein clocks of bacteriophage infections. *Annu. Rev. Microbiol.* **54,** 799–825.

Warren, R. (1980). Modified bases in bacteriophage DNAs. *Annu. Rev. Microbiol.* **34,** 137–158.

Wilcox, R. M., and Fuhrman, J. A. (1994). Bacterial viruses in coastal seawater: Lytic rather than lysogenic production. *Mar. Ecol. Prog. Ser.* **114,** 35–45.

Wommack, K., and Colwell, R. (2000). Virioplankton: Viruses in aquatic ecosystems. *Microbiol. Mol. Biol. Rev.* **64,** 69–114.

Wommack, K. E., Hill, R. T., and Colwell, R. R. (1995). A simple method for the concentration of viruses from natural water samples. *J. Microbiol. Methods* **22,** 57–67.

Xu, M., Wang, I., Deaton, J., and Young, R. F. (2002). The role of holins in bacteriophage lysis. *Abstr. Gen. Meeting Am. Soc. Microbiol.* **102,** 303.

Zhang, T., Breitbart, M., Lee, W. H., Run, J. Q., Wei, C. L., Soh, S. W., Hibberd, M. L., Liu, E. T., Rohwer, F., and Ruan, Y. (2006). RNA viral community in human feces: Prevalence of plant pathogenic viruses. *PLoS Biol.* **4,** 108–118.

Author Index

A

Adams, M. D., 110
Adelberg, E., 3
Aderem, A., 85
Ahmed, A., 183
Akiyama, K., 257
Aldridge, P., 85
Alper, M. D., 186
Altman, M., 260
Alwood, A., 22, 128, 132
Amaladas, N. H., 210, 211, 212, 215, 218
Ames, B., 45
Ames, B. N., 186, 255
Ammerman, J. W., 260
Andersen-Nissen, E., 85
Anderson, E., 47
Anderson, P., 22
Andersson, D. I., 187
Andresen, P., 260, 261, 263
Ara, T., 223
Arber, W., 177
Arsuaga, J., 210
Arvidson, D. N., 228, 229
Au, K. G., 184
Austin, S., 173, 174, 176
Austin, S. J., 210
Ausubel, F., 266
Azam, F., 260, 261, 263, 266
Azam, T. A., 211
Azaro, M. A., 211, 213

B

Baba, M., 223
Baba, T., 223
Badarinarayana, V., 92, 111
Bahrani-Mougeot, F. K., 111
Bailey, J., 83, 84, 85, 126, 128
Baker, T., 191, 192
Baldwin, D. N., 90
Ball, C. A., 96, 103
Ban, N., 74, 83

Baneyx, F., 88
Banks, D. J., 260
Barbour, S. D., 201
Barnes, W. M., 216, 218
Barrett, S. L. R., 85
Bass, S., 227, 228, 229, 230, 238
Bates, D., 210
Becker, E., 85
Beckwith, J., 127, 128, 131, 157
Beckwith, J. R., 127
Bender, J., 11, 12, 53, 66, 159
Bennett, K. A., 228, 229
Benson, N., 227, 229, 230, 238
Benson, N. R., 249, 254, 255
Benton, B. M., 111
Beres, S. B., 260
Berger, I., 173
Bergh, O., 259
Bergman, M. A., 85
Berkelman, T., 185
Berlyn, M., 38
Bertrand, K. P., 203
Bhasin, A., 115, 129, 130
Bibi, E., 74, 83
Binkley, G., 96, 103
Bird, D. F., 259
Bliska, J. B., 211
Blomfield, I. C., 193
Boccard, F., 210
Bochkareva, E. S., 74
Bochner, B., 145, 255
Bolek, P., 71, 74, 83, 84, 85
Bolivar, F., 223
Bond, S., 111
Bonner, D. M., 236
Boocock, M. R., 211
Boos, W., 77
Borsheim, K. Y., 259
Bossi, L., 192, 200
Botstein, D., 11, 22, 25, 30, 38, 39, 40, 43, 96, 103, 111, 228, 250
Bouvier, S., 246
Bovee, D., 22, 128, 132

Subject Index

A

Allele number, bacteria nomenclature, 4
ant, see P22 phage
Antibiotic resistance, nomenclature, 7
Antibiotic-resistant transposon
  applications, 11–12
  delivery systems
    phage, 13
    plasmid, 13
    transposase overexpression *in trans*, 14
  EZ-Tn5 Transposome system, *see* EZ-Tn5
  T-POP, *see* T-POP
  transposase specificity and
    engineering, 12–13
  transposon pools and *cis*
    complementation, 14–17

B

Beta, *see* λ-Red genetic engineering;
  Recombineering

C

Cesium chloride density gradient
  centrifugation, phage particles, 263–264
Cetyl trimethyl ammonium bromide, phage
  DNA extraction, 264–266
Challenge phage system, *see* P22 phage
Chromosome rearrangement, bacteria
  nomenclature, 6–7
Chromosome structure, bacteria
  DNA-binding proteins, 210–211
  λ-integrase probing
    gel-based assay of chromosome fluidity
      cell culture and induction, 215–216
      data analysis, 217–218
      materials, 215
      polyacrylamide gel electrophoresis
        and staining, 217
      polymerase chain reaction, 216–217

genetic engineering using phage
  integrases, 223–224
principles, 212–214
quantitative polymerase chain reaction
  assay of chromosome fluidity
    cell culture, 220
    data analysis, 221–222
    materials, 219–220
    overview, 218–219
    polymerase chain reaction, 220–221
Mu*d*-P22 and chromosomal insertions
  cloning, restriction mapping, and DNA
    sequencing, 256
  DNA hybridization mapping, 256–257
  isolation, 251
  rapid genetic mapping, 253–256
  overview of techniques, 210
  prospects for use in non-*Salmonella*
    species, 257
  site-specific recombination
    studies, 211–212
Conditional allele, bacteria
  nomenclature, 7–8
Conditional mutagenesis
  applications, 43
  temperature-sensitive mutants, 43
CTAB, *see* Cetyl trimethyl ammonium
  bromide

D

DES mutagenesis, *see* Diethylsulfate
  mutagenesis
Diethylsulfate mutagenesis
  incubation conditions and plating, 47–48
  principles, 44
DNA-binding proteins
  bacterial chromosome proteins, 210–211
  P22 challenge phage system
    ant gene expression regulation, 230
    applications, 227–228
    DNA binding selection